I0883211

Literacy for Science

Exploring the Intersection of the Next Generation Science Standards and Common Core for ELA Standards

A Workshop Summary

Holly Rhodes and Michael A. Feder, *Rapporteurs*

Steering Committee on Exploring the Overlap between "Literacy in Science" and the Practice of Obtaining, Evaluating, and Communicating Information

Board on Science Education

Division of Behavioral and Social Sciences and Education

NATIONAL RESEARCH COUNCIL
OF THE NATIONAL ACADEMIES

THE NATIONAL ACADEMIES PRESS
Washington, D.C.
www.nap.edu

THE NATIONAL ACADEMIES PRESS 500 Fifth Street, NW Washington, DC 20001

NOTICE: The project that is the subject of this report was approved by the Governing Board of the National Research Council, whose members are drawn from the councils of the National Academy of Sciences, the National Academy of Engineering, and the Institute of Medicine. The members of the committee responsible for the report were chosen for their special competences and with regard for appropriate balance.

This study was supported by the Contract No. 10001820 between the National Academy of Sciences and the Carnegie Foundation. Any opinions, findings, conclusions, or recommendations expressed in this publication are those of the author(s) and do not necessarily reflect the views of the organizations or agencies that provided support for the project.

International Standard Book Number-13: 978-0-309-30517-4
International Standard Book Number-10: 0-309-30517-9

Additional copies of this report are available from the National Academies Press, 500 Fifth Street, NW, Keck 360, Washington, DC 20001; (800) 624-6242 or (202) 334-3313; http://www.nap.edu.

Printed in the United States of America

Suggested citation: National Research Council. (2014). *Literacy for Science: Exploring the Intersection of the Next Generation Science Standards and Common Core for ELA Standards, A Workshop Summary*, H. Rhodes and M.A. Feder, *Rapporteurs.* Steering Committee on Exploring the Overlap between "Literacy in Science" and the Practice of Obtaining, Evaluating, and Communicating Information. Board on Science Education, Division of Behavioral and Social Sciences and Education. Washington, DC: The National Academies Press.

STEERING COMMITTEE ON EXPLORING THE OVERLAP BETWEEN "LITERACY IN SCIENCE" AND THE PRACTICE OF OBTAINING, EVALUATING, AND COMMUNICATING INFORMATION

P. DAVID PEARSON (*Chair*), Graduate School of Education, University of California, Berkeley
JUAN-CARLOS AGUILAR, Georgia Department of Education
SARAH MICHAELS, Department of Education, Clark University
ELIZABETH BIRR MOJE, School of Education, University of Michigan
SUSAN PIMENTEL, Student Achievement Partners
HELEN QUINN, Emerita, SLAC National Accelerator Laboratory, Stanford University

MICHAEL FEDER, *Study Director*
HEIDI SCHWEINGRUBER, *Deputy Director, Board on Science Education*
REBECCA KRONE, *Senior Program Assistant*

Preface

This summary was prepared by independent rapporteurs. The workshop was designed to explore the intersection between the "Literacy in Science" portions of the Common Core State Standards for English Language Arts and the practices in the Next Generation Science Standards. The views contained in the report are those of individual workshop participants and do not necessarily represent the views of all workshop participants, the planning committee, or the National Research Council (NRC). The planning committee was responsible only for the quality of the agenda and the selection of participants. Neither the workshop nor this summary is intended as a comprehensive review of what is known about the topic. The presentations and discussions were limited by the time available for the workshop.

This workshop summary has been reviewed in draft form by individuals chosen for their diverse perspectives and technical expertise, in accordance with procedures approved by NRC's Report Review Committee. The purpose of this independent review is to provide candid and critical comments that will assist the institution in making its published report as sound as possible and to ensure that the report meets institutional standards for objectivity, evidence, and responsiveness to the charge. The review comments and draft manuscript remain confidential to protect the integrity of the process.

We thank the following individuals for their review of this report: George I. Matsumoto, Monterey Bay Aquarium Research Institute; P. David Pearson, Language and Literacy and Cognition and Development, Graduate School of

Education, University of California, Berkeley; and Sam Shaw, Division of Learning and Instruction, South Dakota Department of Education.

Although the reviewers listed above provided many constructive comments and suggestions, they were not asked to endorse the content of the report nor did they see the final draft of the report before its release. The review of this report was overseen by Carlo Parravano, executive director (retired), Merck Institute for Science Education. Appointed by the NRC, he was responsible for making certain that an independent examination of this report was carried out in accordance with institutional procedures and that all review comments were carefully considered. Responsibility for the final content of this report rests entirely with the authors and the institution.

CONTENTS

1

INTRODUCTION

The recent movement in K-12 education toward common standards in key subjects represents an unprecedented opportunity for improving learning outcomes for all students. These standards initiatives—the *Common Core State Standards for English Language Arts & Literacy in History/Social Studies, Science, and Technical Subjects* (National Governors Association Center for Best Practices and Council of Chief State School Officers, 2010a) and *Common Core State Standards for Mathematics* (National Governors Association Center for Best Practices and Council of Chief State School Officers, 2010b) and the *Next Generation Science Standards: For States, By States* (NGSS; NGSS Lead States, 2013)—are informed by research on learning and teaching and a decade of standards-based education reform. While the standards in English language arts and science have been developed separately, there are areas where the standards intersect directly. One such area of intersection occurs between the "literacy in science" portions of the Common Core State Standards for English Language Arts (CCSS for ELA)[1] and the practices in the NGSS (originally outlined in the National Research Council's (NRC's) *A Framework for K-12 Science Education* [2012]), particularly the practice of "Obtaining, evaluating and communicating information" (Practice 8).[2] Box 1-1 presents the eight NGSS science and engineering practices.

[1]For more information about the CCSS for ELA: Science and Technical Subjects, see http://www.corestandards.org/ELA-Literacy/RST/introduction/ [March 2014].

[2]For the full text of the NGSS Science and Engineering Practices, see Appendix F of the NGSS. Available: http://www.nextgenscience.org/sites/ngss/files/Appendix%20F%20%20Science%20

BOX 1-1

NGSS SCIENCE AND ENGINEERING PRACTICES

1. Asking questions (for science) and defining problems (for engineering).
2. Developing and using models.
3. Planning and carrying out investigations.
4. Analyzing and interpreting data.
5. Using mathematics and computational thinking.
6. Constructing explanations (for science) and designing solutions (for engineering).
7. Engaging in argument from evidence.
8. Obtaining, evaluating, and communicating information.

SOURCE: National Research Council (2012).

The developers of *A Framework for K-12 Science Education* (K-12 framework) opted to focus on "practices" in science, which encompass both knowledge and skills because they are essential for helping students develop a deeper understanding of how knowledge in science is formed. Science practices also help to make knowledge of concepts and ideas more meaningful. As stated in the K-12 framework, "Standards and performance expectations that are aligned to the framework must take into account that students cannot fully understand scientific and engineering ideas without engaging in the practices of science and engineering and the discourses by which such ideas are developed and refined" (National Research Council, 2012, p. 218). Overall, the science and engineering practices do not stand alone but are integrated with content across the grades.

and%20Engineering%20Practices%20in%20the%20NGSS%20-%20FINAL%20060513.pdf [July 2014].

The development of the NGSS, which built upon the K-12 framework, provided further insight into how these science and engineering practices should be applied in the classroom. The NGSS authors used this insight to develop a set of guiding principles that serve as a basis for deeper understanding of the intentions of the practices. Several of these guiding principles are particularly germane to understanding the potential for synergy with the CCSS for ELA literacy in science standards. First, the practices are to encompass each grade band across K-12, growing in complexity and sophistication. In addition, the practices reflect what students should be able to do but do not constitute pedagogy or curriculum. Perhaps most relevant is the recognition that engagement in the science practices is "language intensive and requires students to engage in classroom science discourse" (NGSS Lead States, 2013, Appendix F, p. 3).

Because the CCSS literacy in science standards predated the NGSS, developers of the NGSS worked directly with the CCSS team to identify the connections between the two sets of standards.[3] However, questions about how the two sets of standards can complement each other and can be used in concert to improve students' reading and writing, as well as listening and speaking, to learn science continue to exist.

Throughout the workshop, the presenters explained that implementing the CCSS for ELA and the NGSS faithfully and meaningfully is challenging for several reasons. One primary reason identified is that the objectives for students outlined in the CCSS for ELA and the NGSS, while complementary, reflect disciplinary differences in the kinds of knowledge and skills that are emphasized. In addition, teachers face competing priorities and limited time in the school day in which to accomplish educational goals. Further, the issues to be addressed and the structures of the standards documents vary across grade levels. For example, the literacy in science portion of the CCSS for ELA standards only apply to grades 6-12. However, there are CCSS for ELA elements of the standards in grades K-5 that are potentially relevant to science, but no guidance is provided for teachers as to how to address them in the context of science. In the K-12 framework and NGSS, the intent is that the practice of obtaining and communicating information will be addressed across grades K-12. Another significant challenge is that the organization of schools and the expertise of teachers are not always well-matched to the

[3]Appendix M of the NGSS provides a detailed explanation of connections between the CCSS for ELA literacy in science standards and the NGSS. Available: http://www.nextgenscience.org/sites/ngss/files/Appendix M Connections to the CCSS for Literacy_061213.pdf [July 2014].

shifts in practice that are called for in the standards. While links to the potentially relevant aspects of the CCSS for ELA are included in the NGSS, the specific strategies teachers can use to make these connections will need to be included in instructional materials and curricula that are based on the standards, and they will need professional development that supports the shifts in approach and pedagogy.

Despite these challenges to implementing the CCSS for ELA and the NGSS, research and practice has demonstrated that literacy[4] and science need not compete for priority. Rather, as many workshop participants described, natural synergies exist that benefit both disciplines at the same time to the advantage of both students and teachers. The workshop, summarized here, was designed to explore and provide clear examples of the way in which the CCSS for ELA and the NGSS can work together, and the supports throughout the educational system that make this vision possible.

ORIGIN OF THE WORKSHOP

At the beginning of the workshop, Helen Quinn, Stanford Linear Accelerator Center, Stanford University (emerita) and chair of the Board on Science Education (BOSE), explained that the need for the workshop became evident as BOSE discussed the confusion that still exists among teachers and administrators about how to and who should implement the literacy in science standards of CCSS for ELA and how these standards work with the NGSS. The number and nature of questions from around the country led BOSE to determine that addressing this confusion was a top priority. Therefore, the board initiated plans to develop a workshop to address the need to coordinate the literacy in science aspect of CCSS for ELA with the explicit demands on "obtaining, evaluating and communicating information" as a practice in science instruction. They also wanted to ensure that the workshop addressed concerns from science educators that the new requirement of reading *about* science might prevent the engagement of students *with* science, primarily in elementary grades.

A six-member planning committee worked together over a period of six months to plan a workshop to address these concerns. The committee was composed of individuals with expertise in science and literacy education, classroom discourse, curriculum and professional development, state education policy, and

[4]Literacy is used in this summary to mean the ability to read, write, listen, and communicate information orally.

BOX 1-2

GOALS OF THE WORKSHOP

An ad hoc steering committee will plan and hold a public workshop looking at the intersection between the "Literacy in Science" portions of the Common Core State Standards for English Language Arts (CCSS for ELA) and the practices in the Next Generation Science Standards (originally outlined in NRC's *A Framework for K-12 Science Education*).

The proposed workshop will feature invited presentations and discussion that will

(1) explore the intersections and overlap between the "Literacy in Science" portions of the CCSS for ELA and Practice 8 in the NRC's framework related to "obtaining, evaluating, and communicating information" including consideration of the unique characteristics of communication in science;

(2) consider the complementary roles of English language arts teachers and science teachers as well as the unique challenges and approaches for different grade levels and articulate the knowledge and skills teachers need to support students in developing competence in reading and communicating in science;

(3) consider design options for science and ELA curricula and courses that provide aligned support for students to develop competencies in reading and communicating in science;

(4) discuss the role of district and school administrators in guiding implementation of science and ELA to help ensure alignment.

the intersection of science, literacy, and culture.[5] Committee members planned the agenda and structure of the workshop and identified presenters to help achieve the goals set forth in the charge to the committee, shown in Box 1-2.

The committee planned a two-day workshop with six sessions to meet the goals.[6] Session 1 addressed the nature of literacy in science in the CCSS for ELA and in the NGSS. Session 2 was devoted to a closer examination of the underlying principles involved in literacy for science, as well as the nature of text and discourse in science. Session 2 also featured specific examples of literacy for science enacted in classrooms with joint presentations by researchers and teachers. Session 3 summarized the major issues discussed during the first day of the workshop, and

[5]See Appendix C for information on the steering committee and presenters.
[6]See Appendix A for the workshop agenda.

Session 4 focused on challenges for the education system related to literacy for science. This session featured models of novice teacher preparation and professional development. In addition, presentations also addressed efforts to address the intersection of CCSS for ELA and NGSS on a larger scale. Sessions 5 and 6 focused on identifying and discussing major themes and next steps. Throughout the two days, audience members had numerous opportunities to comment, ask questions, and engage in discussion with one another. The workshop was also Webcast to include remote participants, and audience members and viewers were invited to submit questions and comments on a workshop Webpage.

The workshop was held December 9 and 10, 2013, in Washington, DC, and brought together 53 participants from across the country. In addition, 71 people watched the live Webcast of the workshop for at least 30 minutes, 53 of whom watched for over 2 hours.

ABOUT THIS REPORT

This report presents a summary of the presentations and discussions from the workshop. The chapters do not directly follow the order of the sessions. Instead, the presentations and discussions at the workshop are grouped by topic. Chapter 1 addresses the rationale for the workshop. Chapter 2 includes an overview of the connections between English language arts and science. Chapter 3 describes the presentations and discussion around the nature of science text and talk. Chapter 4 includes examples of literacy for science in practice in the classroom from the perspective of curriculum developers and teachers at the elementary, middle, and high school levels. Chapter 5 addresses models of professional development for both novice teachers and practicing teachers. Chapter 6 presents the strategies and lessons learned from the efforts to scale up support for science education at the network, district, and state level. Finally, Chapter 7 presents themes, as well as some ideas for potential future research and policy, identified by some of the workshop participants. Appendix A is the workshop agenda, and Appendix B is a list of registered workshop participants. Appendix C contains biographical summaries for the steering committee members and workshop speakers. Many presenters prepared background papers to accompany their presentations. These papers are located on the BOSE project Website.[7]

---

[7]The Website is available at http://sites.nationalacademies.org/DBASSE/BOSE/DBASSE_085962 [July 2014].

2

LITERACY FOR SCIENCE IN ENGLISH LANGUAGE ARTS AND SCIENCE STANDARDS

HOW DO CCSS FOR ELA AND NGSS WORK TOGETHER?

In her presentation, Susan Pimentel, a planning committee member, principal of Student Achievement Partners, and one of the developers of the Common Core State Standards (CCSS), addressed the rationale behind the development of the literacy in science and technical subjects standards aspects of CCSS for English Language Arts (CCSS for ELA). First, she said, literacy in science was included in the CCSS for ELA to ensure that science retained a meaningful place in the elementary grades, where reading and mathematics are heavily emphasized, while also recognizing the limited time in the school day. A second motivation for creating literacy in science standards was to ensure that high school students are prepared to access and use science texts,[1] which are often difficult for students to comprehend due to challenging words and grammar, atypical logic structures, and multiple representations. Similarly, Pimentel reported that preparation for postsecondary education was a motivating factor, with about 50 percent of students adequately prepared to handle science and other texts as freshmen in college according to recent data (ACT, 2006). Finally, the literacy in science standards point to the importance of reading and understanding science texts, such as science articles in magazines and newspapers or on the Web, to prepare students

[1]Science texts include a wide range of formal and informal texts, such as textbooks, journal articles, science-focused articles in popular magazines and newspapers, Web content, and notes on science experiments.

for citizenship. Now that states are adopting the Next Generation Science Standards (NGSS), Pimentel emphasized that CCSS for ELA literacy in science standards are "meant to support *not* supplant a state's science standards. They are meant to buttress the teaching and learning of science content."

Next, Pimentel clarified who was intended to be responsible for addressing the CCSS for ELA literacy in science standards. She indicated that teachers in kindergarten through 5th grade generally work across subjects and would address reading, listening, speaking, and writing in science. At this level, teachers should integrate the CCSS for ELA standards into the teaching of core disciplinary ideas, just as they would with social studies, literature, or other disciplines. In grades 6-12, responsibilities differ across the subjects. Middle and high school ELA teachers would not be responsible for meeting the literacy in science standards. She added that she has fielded concerns such as, "ELA teachers now think that they have to teach science." Although ELA teachers do address the use of informational text in their classes, science teachers are responsible for the literacy in science standards at this level. She reaffirmed that the CCSS for ELA standards are merely tools to support the teaching of core disciplinary ideas.

The CCSS for ELA literacy in science standards work in tandem with the NGSS, which address science core ideas, crosscutting concepts, and practices. Pimentel argued that the literacy in science standards are consistent with and affirm the "norms and conventions" of science. For example, she said, the standards call for students to:

- attend to evidence with precision and detail;
- gather, synthesize, and corroborate complex information;
- make and assess arguments orally and in writing;
- make accounts of events and ideas; and
- integrate, translate, and evaluate prose, graphs, charts, and formulas.

Table 2-1 shows how the CCSS for ELA literacy in science standards map onto particular NGSS science and engineering practices.

Brian Reiser of Northwestern University addressed key aspects of the NGSS science and engineering practices and the role for literacy in doing science. In Reiser's view, these practices are a central focus of the NGSS, and they emphasize developing and using science, rather than learning *about* science. In his view, this constitutes a major "evolutionary and revolutionary" shift in science education. The goal is to help students understand why a core idea in science makes sense and how it helps explain phenomena in the world. Reading textbooks about science ideas is insufficient, in Reiser's view, for helping students understand why scientists know what they know

TABLE 2-1 Examples of CCSS for ELA Literacy in Science Practices That Support NGSS Practices

Practice 3: Planning and Carrying Out Investigations			
CCSS for ELA Literacy in Science	Grades 6-8	Grades 9-10	Grades 11-12
Following Complex Processes and Procedures	Follow precisely a multistep procedure when carrying out experiments, taking measurements, or performing technical tasks.	Follow precisely a complex multistep procedure when carrying out experiments, taking measurements, or performing technical tasks, attending to special cases or exceptions defined in the text.	Follow precisely a complex multistep procedure when carrying out experiments, taking measurements, or performing technical tasks; analyze the specific results based on explanations in the text.
Conducting Research	Conduct short research projects to answer a question (including a self-generated question), drawing on several sources and generating additional related, focused questions that allow for multiple avenues of exploration.	Conduct short as well as more sustained research projects to answer a question (including a self-generated question) or solve a problem; narrow or broaden the inquiry when appropriate; synthesize multiple sources on the subject, demonstrating understanding of the subject under investigation.	Conduct short as well as more sustained research projects to answer a question (including a self-generated question) or solve a problem; narrow or broaden the inquiry when appropriate; synthesize multiple sources on the subject, demonstrating understanding of the subject under investigation.
Practice 6: Constructing Explanations and Designing Solutions			
CCSS	Grades 6-8	Grades 9-10	Grades 11-12
Using Textual Evidence and Attending to Detail	Cite specific textual evidence to support analysis of science and technical texts.	Cite specific textual evidence to support analysis of science and technical texts, attending to the precise details of explanations or descriptions.	Cite specific textual evidence to support analysis of science and technical texts, attending to important distinctions the author makes and to any gaps or inconsistencies in the account.
Synthesizing Complex Information	Compare and contrast the information gained from experiments, simulations, video, or multimedia sources with that gained from reading a text on the same topic.	Compare and contrast findings presented in a text to those from other sources (including their own experiments), noting when the findings support or contradict previous explanations or accounts.	Synthesize information from a range of sources (e.g., texts, experiments, simulations) into a coherent understanding of a process, phenomenon, or concept resolving conflicting information when possible.

continued

TABLE 2-1 Continued

Explaining Concepts, Processes, and Procedures	Write informative/ explanatory texts, including the narration of *scientific procedures/ experiments or technical processes*. . . Develop the topic with relevant, well-chosen facts, definitions, concrete details, quotations, or other information and examples.	Write informative/ explanatory texts, including the narration of *scientific procedures/ experiments or technical processes*. . . Develop the topic with well-chosen, relevant, and sufficient facts, extended definitions, concrete details, quotations, or other information and examples appropriate to the audience's knowledge of the topic.	Write informative/explanatory texts, including the narration of *scientific procedures/ experiments or technical processes*. . . Develop the topic thoroughly by selecting the most significant and relevant facts, extended definitions, concrete details, quotations, or other information and examples appropriate to the audience's knowledge of the topic.
Practice 7: Engaging in Argument from Evidence			
CCSS	Grades 6-8	Grades 9-10	Grades 11-12
Making Arguments	Support claim(s) with logical reasoning and relevant, accurate data and evidence that demonstrate an understanding of the topic or text, using credible sources.	Develop claim(s) and counterclaims fairly, supplying data and evidence for each while pointing out the strengths and limitations of both claim(s) and counterclaims in a discipline-appropriate form and in a manner that anticipates the audience's knowledge level and concerns.	Develop claim(s) and counterclaims fairly and thoroughly, supplying the most relevant data and evidence for each while pointing out the strengths and limitations of both claim(s) and counterclaims in a discipline-appropriate form that anticipates the audience's knowledge level, concerns, values, and possible biases.
Assessing Arguments	Distinguish among facts, reasoned judgment based on research findings, and speculation in a text.	Assess the extent to which the reasoning and evidence in a text support the author's claim or a recommendation for solving a scientific or technical problem.	Evaluate the hypotheses, data, analysis, and conclusions in a science or technical text, verifying the data when possible and corroborating or challenging conclusions with other sources of information.

TABLE 2-1 Continued

Practice 8: Obtaining, Evaluating, and Communicating Information			
CCSS	Grades 6-8	Grades 9-10	Grades 11-12
Gathering Relevant Evidence	Gather relevant information from multiple print and digital sources, using search terms effectively; assess the credibility and accuracy of each source; and quote or paraphrase the data and conclusions of others while avoiding plagiarism and following a standard format for citation.	Gather relevant information from multiple authoritative print and digital sources, using advanced searches effectively; assess the usefulness of each source in answering the research question; integrate information into the text selectively to maintain the flow of ideas, avoiding plagiarism and following a standard format for citation.	Gather relevant information from multiple authoritative print and digital sources, using advanced searches effectively; assess the strengths and limitations of each source in terms of the specific task, purpose, and audience; integrate information into the text selectively to maintain the flow of ideas, avoiding plagiarism and overreliance on any one source and following a standard format for citation.
Translating Information from One Form to Another	Integrate quantitative or technical information expressed in words in a text with a version of that information expressed visually (e.g., in a flowchart, diagram, model, graph, or table).	Translate quantitative or technical information expressed in words in a text into visual form (e.g., a table or chart) and translate information expressed visually or mathematically (e.g., in an equation) into words.	Integrate and evaluate multiple sources of information presented in diverse formats and media (e.g., quantitative data, video, multimedia) in order to address a question or solve a problem.

NOTES: CCSS, Common Core State Standards; CCSS for ELA, Common Core State Standards for English Language Arts; NGSS, Next Generation Science Standards.

SOURCE: Adapted from Pimentel (2013).

and how core ideas in science help to explain about the world. The typical practices of reading definitions and explanations, summarizing readings, communicating these readings, and occasionally using this knowledge in investigation do not generally support the sense-making process. Rather, he suggested, using the science practices engages students in using cognitive, social, and language skills in doing the work of science. The use of these practices to build understanding is also in service of building a depth of knowledge about core ideas in science. Ideally, coherence should exist within and across the scientific disciplines to help students build a storyline of explanation that builds on their prior knowledge.

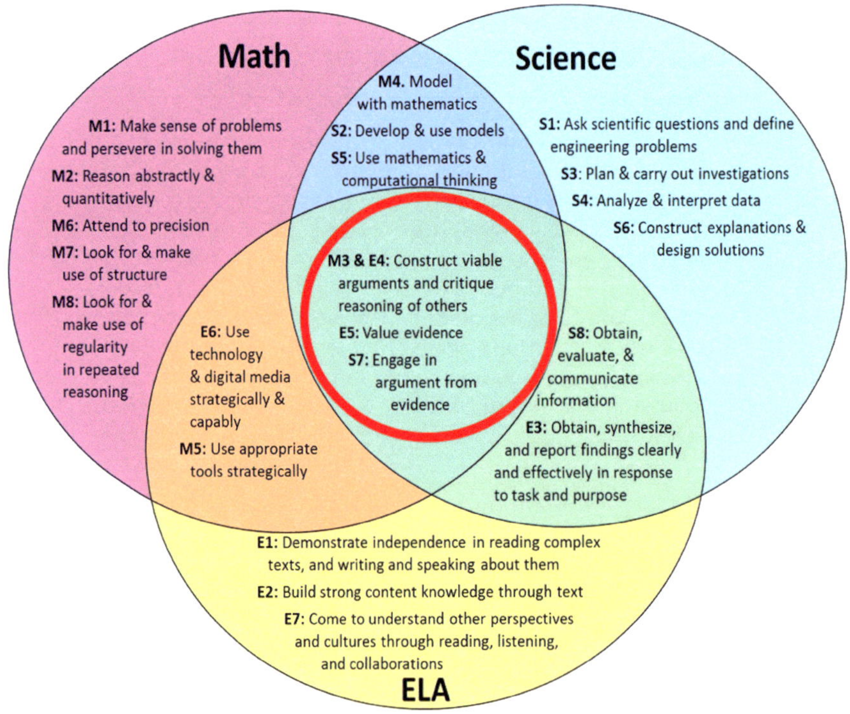

FIGURE 2-1 Overlap between CCSS for ELA, CCSS for mathematics, and NGSS science and engineering practices.
NOTES: CCSS for ELA, Common Core State Standards for English Language Arts; CCSS, Common Core State Standards; NGSS, Next Generation Science Standards.
SOURCE: Adapted from Cheuk (2013) and Stage et al. (2013).

Reiser illustrated his views with two examples that show ways to engage in NGSS practices, using reading, writing, and oral discourse through science practices. In one example, students interacted with a computer model simulation of how an invasive species can alter an ecosystem to construct causal explanations. In a second example, students engaged with a text-based case that also provided them with access to primary data, which they then used to construct explanations of the phenomenon. He pointed out that in both cases, students posed questions, completed readings and investigations to gather information, engaged in argument to refine explanations, and developed causal accounts through language. Reiser emphasized that in both examples, literacy practices play a critical role in helping students "figure things out." Scientific discourse and social interaction are critical to this process of making meaning and developing explanations, he said.

IDENTIFYING THE CHALLENGES AND OPPORTUNITIES

The overlap between science and literacy creates an opportunity and a challenge, explained David Pearson, University of California, Berkeley, and planning committee chair. The opportunity is for synergy between work in various classrooms or subject areas, and the challenge is to maximize that opportunity and avoid conflicts in interpretation and implementation demands between teachers in the different areas.

He said one of the key challenges to achieving the visions of both CCSS for ELA and the NGSS is variation in interpretation. English/language arts educators and science educators are likely to interpret the stated objectives differently. Some of the science practices, such as argumentation, are also misconstrued. In some cases, scientific argumentation becomes a policy debate about a science topic, rather than a sense-making activity. Further, engagement with text is often merely a means to deliver content and not a catalyst for engaging in science practices.

In addition to misunderstanding the standards and their implications, Pearson articulated some of the skepticism that exists between the two disciplines of ELA and science. He shared some of the concerns of both science and language arts educators that he has heard through the course of his work. As he stated, "the first thing I learned when I started working with science educators was that respectable science educators had regarded reading and text as the problem, not the solution to inquiry-based science." For example, he said, round-robin reading of textbooks often replaces the use of science practices and can promote the idea that science consists of a set of facts to be memorized, rather than science as an endeavor. Moreover, many texts are perceived as beyond students' reading levels or include misinformation, according to Pearson. Teachers of language arts, by contrast, have argued that science takes time away from literacy, which is an essential skill, especially in the early grades, whereas science is not always regarded as necessary. Although Pearson has seen improvement in these views, he said he has come to view the essential question as "What benefits can accrue to both literacy and science when we focus on the bridges rather than the barriers between the two?"

Engaging students in reading and understanding texts, as well as helping them develop proficiency in communicating science both orally and in writing, is challenging. Pearson explained that science texts and other modes of communication are typically multimodal (text, diagrams, graphs and charts, equations) and aspire to a level of precision communication of certain details that is unlike that intended in literature. The type of evidence needed is different for science than it is for discussions of interpretations of literature, and thus the strategies for obtaining and evaluating such evidence also differ. This, too, is unfamiliar territory to most teachers and needs further explication as to what it implies for teaching both in the language arts and in the science contexts. He suggested that in order to help students be successful in understanding text and communicating in science, both science and language arts teachers need to be aware of the unique challenges inherent in science communication as well as more general strategies for supporting students' comprehension and expression.

At the middle and high school levels, science teachers likely know the science content they want students to extract from reading a science text, but are often unaware of what it is about that text that makes it difficult for students to comprehend, Pearson explained. As a result, they may focus on strategies such as memorizing vocabulary rather than more sophisticated strategies for text comprehension. Conversely, language arts teachers are generally more aware of the issues of text complexity, but they may be unfamiliar with the specifics of communicating science, unprepared to treat the multimodal aspects of the text, and lack confidence that they can adequately interpret the specifics of the science, particularly those aspects requiring graphs, tables, and equations. Pearson said that at the K-5 level, the same teacher generally teaches both areas and has significant training in teaching reading, but generally little of it focused on science reading. Here the issues are around understanding how effective authentic science learning experiences and discourse can support, rather than compete with, language and literacy development. The major concern that emerged with leading experts in reading instruction is that the explicit call to read *about* science in the CCSS for ELA might be implemented in a way that would prevent the engagement of students *with* science.

SUPPORTING LITERACY FOR SCIENCE

During her presentation, Sarah Michaels, a planning committee member from Clark University, presented a Venn diagram (see Figure 2-1) that depicts a set of four standards shared between the three disciplines of ELA, mathematics, and science, and a set of six standards that address both science and ELA. Central to all three disciplines is placing value on evidence, constructing viable explanations, communicating ideas, engaging in argument based on reasoning, and being able to critique the reasoning of others. However, Michaels and others at the workshop argued that the synergistic relationship between science and literacy indicates that the standards may be integrated even more than this Venn diagram suggests. A number of participants described opportunities for teachers to address literacy and science in mutually beneficial ways. These themes are explained below.

Authentic Reasons to Read and Write

One of the most natural synergies that exists between literacy and science concerns having an authentic purpose to read and write in the classroom, according to David Pearson and other presenters. All reading, writing, and oral language require content. In Pearson's words, "reading and writing are taught best when they reside in disciplinary contexts. . . . Literacy is a set of tools for the acquisition of knowledge and the enhancement of critical thinking rather than a set of goals and ends into themselves." Thus, he suggested, embedding science and literacy together can provide students with meaningful content as they learn to develop the tools for literacy.

This relationship between science and literacy is bidirectional, Pearson explained. Language and literacy skills are critical for communicating in science, and practicing scientists read and write for a number of authentic reasons. For example, scientists read to situate their research in context and acquire new knowledge. They also read so that they are able to replicate procedures and to interpret the data and findings of other scientists. Scientists access reference materials as they plan first-hand investigations of their own. All of these purposes for reading are equally applicable to students of science, Pearson argued.

Having a purpose for reading, writing, and speaking in science commonly means engaging in scientific investigations, according to planning committee member Elizabeth Birr Moje, University of Michigan. She argued that teaching language and literacy for science means that students need to be engaging with the science practices. Otherwise, the language tools have no meaning or value. Using these tools in the service of authentic science, such that the purpose of the learning is evident, may also have an impact on student engagement, she said. However, as Pearson stated earlier, one key bridge between science and literacy is that there are limits to both experiential and text-based ways of learning science. Neither alone is as effective as thoughtfully combining the two.

Pearson and other presenters throughout the workshop emphasized that science involves doing investigations, reading, writing, and talking. According to Pearson, reading, writing, and language do not merely overlap with science; rather, they are interwoven throughout all of the disciplines of school—social studies, mathematics, literature, and science. They do not just have a subset of shared objectives and practices, but rather are inextricably tied to one another, he said.

Gaining in word knowledge means gaining in conceptual knowledge, particularly when paired with using these words in authentic applications. Pearson described many ways to learn new concepts, with new vocabulary as the natural by-product. Learning words in context using repeated experiences and multiple modalities simultaneously builds language skills and conceptual knowledge.

An Inquiry-Based Approach to Learning

Michaels argued that scientific and literacy goals and practices are well-served when they take place within classroom cultures that support public reasoning. In her view, literacy is disciplined reasoning through text and talk. In their presentations, Susan Goldman, University of Illinois, Chicago, and Cynthia Greenleaf, WestEd, noted that such a classroom culture requires a particular stance toward learning and knowledge: that is, a culture that supports engaging in a range of science and engineering practices and values productive struggle toward understanding. Learning is purposefully centered around answering questions. Increasing comfort with ambiguity is another cultural and practice shift that both teachers and students make as they adopt an inquiry-based stance to science learning, as noted by several participants.

According to Pearson, the science practices as laid out in NGSS are comprehension strategies. Comprehension in literacy and inquiry in science are both explicitly focused on making meaning. They share similar goals and strategies, and, he noted, although the nature of evidence differs between the two, the cognitive processes used to reason in science and literacy are fundamentally the same. To illustrate this point, he illustrated that the same strategies and guiding questions can be applied to activities in both science and literacy. In further support of this notion, Pearson pointed out that similar scoring rubrics can even be used across science and literacy using this approach.

The process of making sense of the world, whether through reading or other means, is enhanced when students build on their prior knowledge, according to Brian Reiser. He stressed the importance of helping students build a depth of knowledge. To accomplish this, the curriculum they experience needs to build a coherent storyline, Reiser suggested. This means avoiding the common approach of moving from topic to topic and learning sets of disjointed facts.

Learning How to Communicate about Ideas: Discourse Communities

With science framed around learning core concepts using key science practices, Pearson remarked that at their most basic level, words are merely the labels for concepts and ideas, and the ways the people name their knowledge and the processes they use to learn about the world. Learning the language of science entails learning an array of words that can be organized into conceptual networks. Science involves using particular language to describe, predict, synthesize, and argue, based on certain norms and conventions that differ from those used in everyday life, according to Pearson and Moje. Therefore, understanding scientific concepts is not only experiential but also of necessity about language, as Moje summarized. She said, "The natural sciences are discourse communities or cultures, dependent on oral and written language for producing, communicating, and evaluating knowledge. [Thus], learning science is as much about learning how to use the language of science—both oral and written—as it is about learning concepts."

Moje indicated that to teach the language of science, teachers must examine what words, phrases, and symbols mean in a given subject area or discipline and understand the ways that people use language in the discipline. Then, they also must evaluate why, when, and how these "ways" are useful, as well as why, when, and how they are *not* useful in order to help students learn to understand and use the language of science. This means moving away from teaching vocabulary out of context. Instead, according to Moje, teachers must engage in scientific practices, then elicit and engineer necessary knowledge, skills, and practices in science that can be used to make meaning. All of these practices around literacy for science require knowledge of content, pedagogy, science practices, and texts, as well as sufficient time and skills, Moje said.

Many presenters noted that the public interchange of ideas is at the heart of science. This requires an ability to use academic and disciplinary language to communicate ideas and to understand the reasoning of others through listening, speaking, reading, and writing. As Michaels stated, for students to do the scientific practices described in NGSS, they "have to participate in these practices with others primarily through talk, joint attention, and shared activity." One purpose of this discourse is to reveal student thinking. She added that "if we are serious about promoting the thinking practices at the heart of the CCSS for ELA and NGSS documents, we need to see a change in terms of the kind of classroom talk that teachers facilitate," a vision that she sees as achievable by giving teachers "talk moves" that they can use in the moment during discussions, along with challenging content to discuss and video exemplars.

The language of scientific argumentation as a particular approach to communicating about ideas was a focus of many presenters. One presenter noted that educators explicitly teach students how to make sense of the way scientists read, write, and talk, and they help convey those conventions to students. Others noted that to argue in science is distinctly different from the negative connotation it carries in everyday use. In science, argument involves analysis of a line of thinking and evaluation of evidence to develop an explanatory account. Argument in science also involves hearing constructive feedback from others about the ideas presented, according to Reiser. As he noted, this involves weaving claims together in a causal account, and pointing to the evidence through language that connects directly to data that students have to interpret. Often students can engage in these practices through access to primary data and text sources, particularly for problems that do not lend themselves to first-hand investigation.

As students engage in developing explanatory causal models of scientific phenomena, discussion focuses on building consensus around an explanatory model and not on who has the right answer, Reiser and others pointed out. Thus, teachers shift their approach from Initiation-Response-Evaluation to Claims, Evidence, and Reasoning. A key challenge for teachers across all domains, as noted by Michaels and other participants, is helping and allowing students to grapple with ideas they encounter in this process. This means allowing the answer to reside in the evidence, and not with the teacher. As noted by more than one participant, the evidence that supports an explanatory model becomes the authority but also challenges teachers to resist the urge to tell students the answers and to avoid having students try to guess what answer the teacher wants.

Pearson summarized the literacy skills that students need to have to be able to engage in this type of discourse with "five C's." First, students must be able to *comprehend* the various types of science texts that they read. Second, students should be able to *critique* and evaluate claims. Third, they can *construct* explanations for phenomena using critical thinking and reasoning. Fourth, students need to be able to *compose* their ideas orally and in writing to share with others. Fifth, students need a range of *communication* skills to engage with others throughout the scientific process.

Michaels argued that individuals involved with science education should "join forces with our colleagues in other disciplines, who view literacy as reasoning." Several presenters emphasized that learning how to engage in scientific explorations, understand the nature of evidence, and use reasoning to evaluate claims and arguments is not just about science but is a life skill.

3

THE LANGUAGE OF SCIENCE TEXT AND TALK

Text and talk in the science classroom constitute two of the primary vehicles by which students gain knowledge and make meaning. Yet, both involve the unique language of science. Presentations focused on science texts addressed the importance of text for engaging in investigations; the functions and challenges of its specific forms; and ways that teachers can help students unpack science texts, gain knowledge, and express their own ideas through writing. In addition to reading and writing science texts, students engage in science talk that includes among various purposes: reporting, explaining, questioning, and arguing. Scientific discourse was discussed in a number of presentations representing different ways of thinking about this topic—a rationale for thinking of discourse in terms of reasoning skills and understanding discourse in terms of the language skills it requires of students. Both presentations addressed what it takes and what it means to engage in oral scientific discourse in a classroom structured around authentic design and science.

A number of common themes emerged from the presentations that commented on scientific discourse:

- Reading, writing, and well-structured talk are all authentic aspects of engaging in the sense-making process in science classrooms.
- Science texts come in many forms and have unique and challenging words, grammar, patterns, and representations.

- Teachers of science have an important role to play in helping their students become scientifically literate, and they need certain knowledge, skills, and strategies to do this.
- Science reading, writing, and discourse are uniquely complex, explicit, and precise and require students to use particular sets of receptive and productive language skills.
- Engaging with science texts and productive oral discourse requires teachers to spend time allowing their students to grapple with challenging texts and ideas.

THE IMPORTANCE OF SCIENCE TEXTS FOR "DOING" SCIENCE

Jonathan Osborne, Stanford University, argued that science is primarily about ideas or concepts, and that fully understanding scientific concepts requires engaging in reading, writing, talking, and drawing, in addition to participating in hands-on experiences. Further, each of these activities is essential to the eight scientific practices enumerated in the Next Generation Science Standards (NGSS), particularly Practice 8: Obtaining, evaluating, and communicating information. Science texts not only convey information about particular content and scientific phenomena, but also convey ideas about the central activities of science. According to Catherine O'Connor, Boston University, the language of science texts expresses how scientists expand and refine their ideas and find new ways to solve persistent problems. She said the language requires readers to master not only basic and intermediate literacy skill, but also to understand the intricate, discipline specific literacy skills. Science texts, especially textbooks and journal articles, are constructed using complex sentence structures that can be particularly challenging for students. Annemarie Palincsar of University of Michigan similarly noted in her presentation:

Science texts are a good example of the challenging text to which the framers of the CCSS refer. These texts often present information that is conceptually rich, but also conceptually dense and abstract; they use terminology that is unfamiliar to many students; and they present explanations using language in ways that students do not encounter in their everyday uses of language or in their reading of fictional and narrative text (Palincsar, 2013, p. 10).

O'Connor indicated that science texts are unique in a number of ways that require science teachers to assist their students in learning how to use and construct meaning from them. First, as described in Chapter 2, science texts are usually multimodal, including prose, as well as diagrams, charts, mathematics, or other types of visual representations. Moreover, these elements cannot be understood on their own and are only understood in relation to one another. Second, science texts are often lexically dense, when compared with texts in other subjects, reflecting the conventions of scientific writing. Third, the sciences have particular discipline-specific terms and concepts that require the expertise of teachers in that discipline.

Types of Science Text

Throughout the workshop, presenters noted that students may encounter science writing through their textbooks, but also through a variety of outside sources, such as science journals, popular magazines, or Web-based content. Even science textbooks written at an appropriate reading level contain particular concepts, language, and constructions that may require teacher scaffolding for comprehension. Publications other than textbooks can be useful, but may contain language or sentence construction that is unfamiliar to students.

Within both textbooks and other sources, science writing is designed to convey meaning for particular purposes. Mary Schleppegrell, University of Michigan, stated that science texts often fall into certain genres based on the purpose of that text, and these various types may be found within a single source, such as a textbook, that students would encounter. She explained these genres include definition, explanation, recount/procedure, and argument, and each serves a particular purpose in conveying scientific information:

- Description: To define something or tell what it is like
- Explanation: To tell how or why something works or is as it is
- Recount/procedure: To tell about what happened or what someone did
- Argument: To persuade that something should be done

Schleppegrell pointed out explanation texts are typically characterized by technicality, dense text, development of information from phrase to phrase, meaningful connections between phrases, and words and phrases that convey author perspective. Each genre contains identifiable patterns of word choice that prove useful in deciphering their intent and meaning.

In her earlier presentation, O'Connor focused more specifically on the grammatical construction of science texts, describing her work examining sources of science information beyond traditional textbooks, such as science journals or popular magazines with science topics. She explained the complex grammatical constructions in these texts tend to fall into families of certain types of constructions. One type is the "Comparatives." These types of sentences include overt comparative language, such as "smaller than" or "fewer than," but also covert comparatives, in which a comparison is implied, but not made explicit. A second family of grammatical constructions is the "Conditionals." These are sentences that indicate situations or conditions necessary for a second part of the statement to be true; for example, "Had it survived to adulthood, it would have been 6 feet long." According to O'Connor, "Complex constructions like these are important: They signal logic and purpose, and temporal and quantitative relations, among other things."

One particular form of the conditional construction is the "counterfactual conditional," O'Connor said. This type of sentence may begin with a phrase, such as "were it not for . . .," a phrase which is formulaic and fixed in its construction. These sentences can be difficult to interpret, but can also be rephrased using more everyday language, such as "If _______ hadn't been . . ." or "If _____ was not. . . ." O'Connor stated, "So understanding complex constructions like the comparative (whether covert or overt) and the conditional (whether counterfactual or not) is part of learning to *obtain, evaluate, and communicate information.*"

The Role of Teachers in Using Science Texts

Osborne focused on the important role that science teachers play in helping their students to become scientifically literate. He suggested that science teachers are responsible for focusing on discipline-specific literacy and need specific knowledge and teaching strategies to achieve it, stating:

> *The basic point that I think I would like science teachers to have is that if you are talking about making people scientifically literate, literacy means what it means. . . . There is a fundamental sense of literacy . . . which is the ability to construe meaning from text and to construct meaning with text, as well. . . . The job of a teacher is to help people or students learn how to construct that particular meaning . . . and the way in which that is done is different depending on which discipline you happen to be in.*

Osborne identified key pedagogical content and strategies that science teachers need related to helping students understand the language of science text. First, he stressed that knowledge of pedagogical strategies matters for student outcomes, citing the work of Sadler and his colleagues (2013). Teachers need knowledge of instructional and diagnostic tasks, knowledge of student cognition, and common

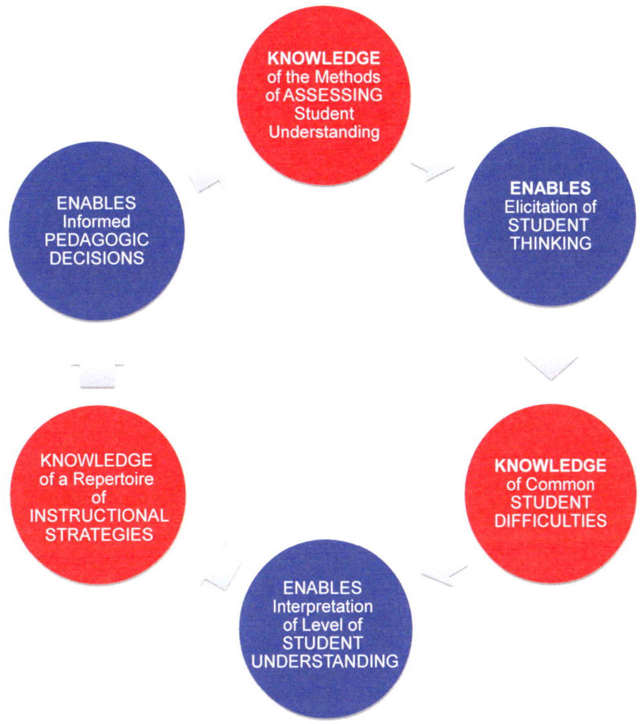

FIGURE 3-1 How knowledge of pedagogy supports teaching practices.
SOURCE: Osborne (2013).

difficulties that students often manifest, as well as knowledge of a range of explanations for and ways of representing and communicating scientific ideas. He also argued that teachers need a repertoire of instructional strategies, including how to activate prior knowledge, promote comprehension, and build recall abilities. As illustrated in Figure 3-1, each aspect of teacher knowledge enables the application of practices that continue to inform teacher knowledge in an ongoing cycle.

Osborne explained some specific methods that teachers can use with science texts to promote recall of information: "Anticipation Guides" to identify and build on prior knowledge (Smith, 1978), Directed Activities Related to Text (DARTs)[1] to promote comprehension, and the Frayer model (Frayer et al., 1969) and Cornell notes (Pauk et al., 2008). All emphasize the need for students to be reflective when they engage with science texts. Further, Osborne stressed that teachers can help students become critical readers of science texts, but cautioned

[1]For more information on DARTs, see https://www.teachingenglish.org.uk/article/interacting-texts-directed-activities-related-texts-darts [March 2014].

that this guidance requires teachers to have sufficient expertise and subject-matter knowledge, a current challenge that requires realistic expectations. He ended his remarks emphasizing that science is about ideas that have to be communicated in written and oral language practices specific to the discipline. He said these overarching principles frame the need for a set of core knowledge and strategies that teachers can use to help their students succeed.

O'Connor emphasized that teachers need not focus on teaching grammar, but rather devote class time to allowing students to grapple with the meaning of complex sentences. She shared strategies that are being used in Lily Wong Fillmore's work with English language learners. In Fillmore's work, teachers can have students paraphrase a text using their own words, by encouraging them to "look up" to the storyline and "look down" into the details. Although teachers may be tempted to summarize and present the meaning of a text to students when it involves challenging language, allowing students to dig deeper into science texts can impart particular benefits to them, according to O'Connor. Namely, the overarching messages of these texts impart knowledge about the work of science/engineering, the texts provide facts and arguments needed to support these story lines, and finally, the contents of these texts help structure teachers' efforts to support students grappling with complex language. Paraphrasing complex text and the resulting discussions can be key parts of the meaning-making process in science.

Mary Schleppegrell and Annemarie Palincsar of the University of Michigan expanded upon ways that elementary teachers can help students understand lexically dense texts by analyzing the language and patterns that authors use in science writing. Several bodies of literature have informed the development of the curriculum *Functional Grammar Analysis*, including systemic functional linguistic theory (Halliday, 1994; Schleppegrell, 2001, 2004), as well as theories that emphasize the linkages between form and meaning in reading situated within a sociocultural context (August and Shanahan, 2008; García and Cuellar, 2006; Goldman and Rakestraw, 2000; Graesser et al., 2003; Sweet and Snow, 2003; Vygotsky, 1986). Further, Schleppegrell and Palincsar have drawn upon the work of Putnam and Borko (2000), which has shown the importance of using teachers' own classrooms as powerful contexts for their learning.

Informed by these theories of linguistics and learning, *Functional Grammar Analysis* is a curriculum for elementary school teachers to be used in the context of language arts, Palincsar explained. It is one tool to help teachers use text and learn to read with students in ways that engage students in thinking about scientific concepts. This is accomplished by focusing detailed attention on the language the author chose and how these choices build meaning. The curriculum involves interactive reading and discussion of text, first-hand investigations, demonstrations of phenomena, and support for writing about the phenomena.

Palincsar noted the effectiveness of this curriculum with a diverse group of students has been supported through research (Palincsar et al., 2013). She described a study in which 26 teachers from grades 2 through 5 participated with 12 coaches/resource teachers to implement this curriculum in their classrooms in Dearborn, Michigan, home to the largest population of Arab Americans in the United States. Over 90 percent of children in these participating classrooms were bilingual, with a high proportion classified as English language learners, and over 90 percent of the students in the schools in their research qualified for free and reduced-cost lunch. After using *Functional Grammar Analysis*, which constituted the only science teaching most students received, students' science content knowledge increased. In addition, analysis of student writing showed an increase, on average, of five idea units from the pre to postwriting assessment, an increase in the range of ideas children included in their explanations, and more use of writing with connectors and author attitude.

With *Functional Grammar Analysis*, Palincsar explained, teachers address the technical nature of science texts by helping students identify certain patterns in the language. For example, a paragraph that includes a series of sentences with "being" or "having" verbs tends to convey a definition. Similarly, students are guided to look for sentences that include the phrase "is called" to further build on definitions. Teachers can also point out the ways in which even the word "or" can be used to indicate a definition. Using these tools, she said, teachers help students see how meanings of technical words become clearer as the text evolves from beginning to end.

A focus on "doing" processes rather than "being" or "having" is characteristic of explanatory text, Palincsar explained. Teachers encourage students to identify meaningful "chunks" of text, purposefully using the words *participants* and *processes*, rather than on nouns and verbs, to emphasize conceptual understanding over parts of speech. In these texts in which the purpose is to describe how something happens, the flow of ideas often follows an identifiable pattern. A concept named at the end of one phrase is used at the beginning of the next, and ideas build upon one another. Connections between phrases also have particular meaning in science texts. They can convey present time, cause, condition, contrast, or other linkages. Examples of these various text patterns are shown in Box 3-1.

BOX 3-1

EXAMPLES OF SCIENCE TEXT PATTERNS: DEFINITION, EXPLANATION, CONJUNCTIONS, AND ATTITUDES

Definition: Looking for "having" or "being" words

When a material ***has*** *electrons that are able to move very freely, it conducts electricity. We* ***call it*** *a conductor. Most metals are good conductors.*

Explanation: Looking for "doing" verbs

The electric current ***provides*** *energy that makes things run. The electrons* ***flow*** *through wires that are made of metal (conductors) and covered in plastic (an insulator).*

Conjunctions: Looking for words that explain how ideas are connected

The energy of the electrons is converted to heat or light ***as*** *the electrons make resisters run.*

Attitudes: Looking for words that express the author's perspective

Likelihood—*could, might, perhaps*

Attitude—*unfortunately, surprisingly*

Connectors that convey perspective—*in fact, but, although*

SOURCE: Schleppegrell and Palinscar (2013).

Last, Schleppegrell indicated that although science texts can seem objective and impersonal, author word choice conveys a perspective on a range of ideas. Authors choose words to convey their ideas about certainty or likelihood or their attitude about a concept. Connecting words, such as "but," "although," or "in fact," can convey author perspective as well. Examining these texts for author perspective is part of the process of identifying claims an author makes and the evidence used to support that claim, which encourages students to be critical readers of text. According to Palincsar (2013, p. 14), "Students who have been supported to learn the scientific practices identified in the NGSS are equipped to bring such a critical stance to text."

THE IMPORTANCE OF SCIENCE TALK IN THE CLASSROOM

Bringing a critical stance toward ideas based on reasoning and learning to engage in scientific argumentation are key elements of scientific discourse, according to two presenters who focused their remarks specifically on the importance of scientific discourse in the classroom. Sarah Michaels, Clark University, addressed the centrality of discourse as part of the social nature of science. Later in the workshop, Okhee Lee of New York University shared her views on this topic, providing an initial framework for considering the analytic, receptive, and productive language functions that scientific discourse in the classroom require.

Science Talk as Public Reasoning

Michaels argued that literacy is "disciplined reasoning through text and talk" and that these reasoning practices have to be enacted. That is, students learn how to reason—constructing, engaging in, and critiquing arguments based on evidence—primarily through talk, attention, and shared activities with others. These social activities can include writing in addition to talk, but have as their ultimate aim to make student thinking public and available to the other members of the community. Thus, she said, the challenge becomes creating classroom environments that support this type of structured social interaction and public reasoning. In fact, Michaels argued that all scientific practices involve these public reasoning practices. In her view, "well-structured talk—discussion or guided, scaffolded reasoning talk—will have to become the new foundation for all of the practices in the Common Core and NGSS."

As Michaels noted, very little of this type of discourse is currently happening in classrooms today, and typically, teachers use an Initiation-Response-Evaluation approach to classroom discussions (Cazden and Mehan, 1989). Such discussions involve the teacher asking a question that generally has one right answer, seeking a typically short response from a student, and then evaluating the correctness of that response. She said this type of discourse is prevalent for a number of reasons. First, most teachers experienced this form of interaction themselves as students. Second, the discussion can be fast-paced, enabling the teacher to cover a lot of material in a relatively short amount of time. Teachers also retain control over the discussion. Although this approach is useful for quick evaluations and checking student knowledge, significant changes are needed to move from this approach to one that promotes reasoning, according to Michaels.

To promote a culture shift to discussions centered around reasoning in classrooms, Michaels stated that teachers need particular forms of support beyond the guidance they have received in the past. Providing teachers with broad "rules of thumb" regarding how to guide classroom discussions, such as to ask higher order questions and avoid those with known answers, fails to provide the level of guidance that proves to be useful. In contrast, Michaels suggested that teachers benefit much more from learning specific "talk moves." These moment-to-moment strategies are designed to help students learn how to explain their own reasoning to others and build on the thinking of others. Successful strategies to help teachers share three common elements: (1) a framework of shared goals and a set of talk moves and strategies focused on accomplishing those goals; (2) challenging and coherent content to discuss; and (3) collections of video examples of scientific discourse. Regarding this third element, Michaels stated, "Teachers can't do what they can't even imagine."

According to Michaels, centering classrooms on and providing science teachers with support in reasoning-focused classroom talk is a "high-leverage" strategy. The impact of reasoning-focused classrooms on student thinking and learning can be significant because teachers must carefully consider content, learning goals and expectations, the cognitive demands of the task, and the knowledge possessed, perceived, and to be learned by students. Because of their impact and centrality to learning the scientific practices, Michaels argued that they should be the center of science teaching.

Understanding the Nature of Science Talk: Analytic and Language Functions

Okhee Lee examined the CCSS and NGSS to identify the extent to which science discourse is emphasized and how it is described. Overall, she found that while both point to an important role for talk as a scientific practice and as a way to learn content, talk is far less emphasized than is writing. Discourse appeared to lack a clear definition, she said. Beyond the standards, she noted that students have varying levels of exposure to language used in science and teachers generally do not model the language practices of science. Student writing often mirrors the way they speak.

Lee observed that despite these challenges, oral discourse in the classroom can benefit both science and language development. She argued that science learning is based on experience. Experience, in turn, is essential for the development of oral language. The development of oral language supports written language and is critical to the construction of meaning. She added that these linkages between experience, oral and written language, and meaning-making are not one-directional, but rather are linked together in a more complex feedback loop. Teachers can scaffold the way to use language in scientific oral discourse to support both science learning and writing. These oral and written skills can ultimately prove useful in other subjects, she noted.

Lee stated that engagement in the science classroom requires students to perform particular receptive and productive language tasks. They must both comprehend oral and written language, as well as communicate their ideas through talk and writing. Each of the NGSS science and engineering practices requires particular sets of these receptive and productive language tasks, along with a set of analytic tasks. Examples of the tasks needed for NGSS Practice 7: Engage in Argument from Evidence are found in the English Language Proficiency Development Framework (Council of Chief State School Officers, 2012; Table 6, pp. 29-30). Further, Lee described, students' receptive and productive oral and written language tasks can be broken down into various modalities, like whole classroom or small group, multiple ways of speaking or registers, and examples of those registers (Council of Chief State School Officers, 2012; Table 9, p. 35).

Finally, Lee explained that in a classroom centered on scientific investigations, language should be precise, explicit, and complex. Just as with text, science talk involves using particular words with particular meanings beyond those used in everyday speech. Science talk involves detailed reporting and/or explaining one's thinking about ideas and actions in clear terms. In addition, being able to describe relationships and connections using oral and written language demands a level of complexity in terminology as well as in ways of putting ideas together. As a way of examining the extent to which classroom talk in science reflects these language qualities, Lee presented a series of questions that could help determine whether science discourse goals were being met, as shown in Figure 3-2. Overall, she emphasized that science classrooms are important language learning environments and that oral discourse is a key element of this language environment. This oral discourse is important to the sense-making process in science and requires teacher support for students but also professional support for teachers to achieve success for all students in science.

Precision

- Does the discourse use discipline-specific terms appropriately?
- Is the discourse exact enough to communicate nuanced meaning?

Explicitness

- Would the audience understand the discourse without context?
- Could someone who is not in the classroom understand the discourse?
- Does the student appropriately use logical connectors (e.g., because, since, therefore, so) to be explicit about relationships between ideas?

Complexity

- Does the student explain why?
- Does the student provide evidence to support a claim(s)?
- Does the student communicate about relationships between concepts?

FIGURE 3-2 Questions to help determine whether science discourse goals are being met.
NOTE: Brianna Avenia-Tapper came up with the conceptualization of the ideas expressed in the figure.
SOURCE: Lee and Llosa (2011-2015).

DISCUSSION

Audience members and panelists took part in a discussion about how to use these practices, particularly those involving challenging science texts, with varying reading levels present in many classes. Osborne shared that teachers need a way to know student reading level and then to engage in science reading while bolstering student confidence in their reading competence, and to differentiate instruction accordingly. However, Michaels and O'Connor cautioned against diluting the content and complexity that students encounter. Helen Quinn suggested that helping students grapple with complex sentences and equipping them with tools are both parts of helping students become better readers. Affirming this approach, O'Connor stated that much of her work occurs in schools where many students read below grade level. She urged workshop participants not to be afraid to help their students grapple with difficult language, and to use discourse to ask them what they think.

4

WEAVING SCIENCE AND LITERACY TOGETHER ACROSS THE GRADES: EXEMPLARS

Weaving science and literacy instruction together successfully is not just a theoretical ideal. Across various levels of K-12 education, curricula that address literacy for science have been developed, implemented, and evaluated. Developers of four different curricula—*Seeds of Science/Roots of Reading*, *Science IDEAS*, *Investigating and Questioning our World through Science and Technology (IQWST)*, and *Project READi*—made presentations at the workshop that described the development process, key elements and defining features, and evaluation results. Two of these presentations were paired with a presentation from a teacher who had implemented or was implementing the curriculum in her classroom. These complementary presentations provided the audience with detailed and specific examples of teacher strategies, student work process and products, as well as teachers' reflections on their experiences.

The following themes emerged from the presentations:

- Curricula that successfully integrate science and literacy exist and are being implemented at various levels across K-12.
- Teachers can effectively implement these practices.
- Student outcomes in both science and literacy can improve as a result of using these curricula including diverse student populations.
- In cases with longitudinal follow-up, benefits of the curriculum can persist for years beyond use of the curricula.

ELEMENTARY SCHOOL EXEMPLAR 1: SEEDS OF SCIENCE/ROOTS OF READING

Jacqueline Barber, Lawrence Hall of Science at University of California, Berkeley, shared her insights into developing *Seeds of Science/Roots of Reading*[1] in collaboration with David Pearson. From a science perspective, Barber was motivated by a desire to ensure that science had more time and retained a meaningful place in the elementary curriculum. She shared that Pearson considered "literacy a domain in search of content." She commented that their common interests and their divergent backgrounds and perspectives led to a collaboration with some tension and debate, but one that was ultimately productive. According to Barber, this partnership led to solutions that could meet the needs of both those focused on science education and those focused on literacy, while remaining mindful of the need to address multiple content areas efficiently in the limited time of the school day. Amid the pressures of high-stakes testing, she said literacy remains the primary focus of the school day, particularly in the early grades, so a need exists for a solution integrating science and literacy, which was reflected in focus group testing with early elementary teachers. During the focus group testing, statements such as "if this science unit can't do work for us in reading, writing, listening, and speaking, we really just can't do it" were common among participants. Overall, Barber and Pearson's collaboration resulted in a curriculum that is, according to Barber, "100 percent science and 50 percent literacy."

While science educators often value more hands-on investigation over reading and writing, Barber affirmed the importance of investigations that include text, not only for its literacy benefits, but also because the use of text sources is an important element of science. Practicing scientists write and talk to each other to share knowledge, to learn from each other, to access reference materials, and to engage in disagreements.

The purpose of *Seeds of Science/Roots of Reading*, according to Barber, is to promote students' sense-making of the natural world, with students ultimately developing more accurate explanations over time using four basic elements: Do, Talk, Read, Write. The curriculum combines these four practices through tightly paired first- and second-hand investigations focused on answering key questions. First-hand investigations consist of hands-on investigations of scientific phenomena, whereas second-hand investigations include using texts to help answer the same fundamental question as the first-hand investigations are designed to

[1]See http://www.scienceandliteracy.org/ [March 2014].

answer. She stated that neither first-hand nor second-hand experiences in and of themselves constitute science, unless they have a purpose. However, when paired together in service of answering a common scientific question, these experiences can allow for greater depth of knowledge and understanding for students. Thus, written and oral discourse are embedded in the center of all first- and second-hand experiences with the ultimate goal that students will be able to construct explanations, gather evidence, and make arguments they can support about key ideas in science. These pairings of first- and second-hand experiences are carefully sequenced to form a curricular unit.

As teachers begin to implement the curriculum in their classrooms, they provide students with explicit instruction, and then support the students through scaffolding faded over time. For example, after students are provided with specific instruction in how to write explanations and construct arguments, they are then given prompts for their writing, such as, "I think this because . . .," "My evidence is . . .," or "Why is that the best explanation?" Students learn the language of science, including claims, evidence, and reasoning through the curriculum, Barber said.

Sherrie Roland, a teacher at Grafton Village Elementary School in Stafford County, Virginia, shared her experiences in implementing *Seeds of Science/Roots of Reading*. She encouraged other practitioners to combine science and literacy instruction. Describing how her classroom looks to outside visitors, she stated, "They don't know if I'm doing reading in my classroom or science in my classroom, and that's how I like it because the kids are just learning and they are learning well." As evidence of the effectiveness of this approach, she also added that her students have higher literacy, mathematics, and science scores than their counterparts not using this approach.

Further illustrating what each of the four basic practices looks like in a *Seeds of Science/Roots of Reading* classroom, Roland provided descriptions of each. Reading can occur individually, paired with a teacher, or as a collaborative activity between two or more students. She added that she must tailor her support for reading based on students' reading levels and individual needs. However, all students in her class, including English language learners and students with special needs, work with the same content. As students learn skills in scientific discourse, they learn to talk to each other using the language of argumentation, skills that she has found to be applicable across other subjects. In writing, her students make frequent use of "sticky notes" as they talk and read to make predictions and express and refine their ideas. They keep notebooks, making their entries accurate, big, colorful, and detailed. At the beginning of a unit, students write a "Line of Learning" to express what they think a concept means and then return to this writing at the end of a unit, revising their ideas with their new knowledge. Roland also maintains a classroom blog as another form of writing that parents can see. Overall, she emphasized that all of her students value investigating and making sense of their world.

Barber shared that evaluation of *Seeds of Science/Roots of Reading* across classrooms indicates that combining science and literacy instruction in this manner was not only efficient use of time in the school day but also effective in improving student outcomes. In experimental studies of units taught in grades 2-5, comparing performance of students in *Seeds of Science/Roots of Reading* classrooms with performance of students in classrooms with comparable content being taught using the "business as usual curriculum," she reported researchers found that students in *Seeds of Science/Roots of Reading* classrooms always had higher scores on measures of science conceptual knowledge and vocabulary than did control students. In addition, they always performed equivalently or higher than control students on measures of science reading comprehension and science writing. *Seeds of Science/Roots of Reading* classrooms also had more student-to-student talk. Overall, the evaluation revealed gains in science measures with effect sizes as great at .61 compared to control classrooms after a single 8- to 10-week unit of instruction, with no losses in literacy scores despite less explicit focus directly on literacy skills.

ELEMENTARY SCHOOL EXEMPLAR 2: SCIENCE IDEAS

Nancy Romance of Florida Atlantic University described *Science IDEAS*,[2] a curriculum for older elementary school students to teach literacy within science. In the classrooms implementing this curriculum, the language arts block was replaced with *Science IDEAS* and literature filled the half-hour block of time previously devoted to science. She noted that development and initial implementation of this approach occurred in the late 1980s prior to *No Child Left Behind,* after which replacing the language arts block would have been more challenging. The curriculum was initially implemented in several South Florida 4th-grade classrooms, and later also implemented in classrooms targeting drop-out prevention and at-risk students over multiple years. She and her colleague Michael Vitale conducted longitudinal research to measure the effects of the curriculum.

[2]For additional information about Science IDEAS, see http://sites.nationalacademies.org/DBASSE/BOSE/DBASSE_085962 [March 2014].

Several bodies of research influenced the development of *Science IDEAS*, according to Romance. Bransford's work on expertise and how experts operate (National Research Council, 1999) showed the importance of well-organized knowledge and being able to access and apply this knowledge with automaticity. Romance and Vitale also drew upon work on problem solving and application, including how knowledge is transformed from declarative to procedural (Anderson, 1987). Work in the area of knowledge-based instruction and intelligent tutoring systems (Brown, 1989) underscored the importance of structure and coherence of knowledge and instruction. Romance indicated that theory and research on reading comprehension also informed the curriculum.

Romance described the approach and key features of *Science IDEAS*. Science concepts are the focal point of the curriculum with activities, such as reading comprehension, writing, and application, stemming from the focus on the science idea. A key component of this approach is the use of propositional concept maps that show how ideas in science are connected to one another. Figure 4-1 presents an example of a *Science IDEAS* concept map. Teachers help students develop these concept maps over the course of a unit, as the students gain more information based on their observations, reading, and other supporting activities. Supporting activities begin with activating prior knowledge, and then move to identifying real-world examples of the phenomenon. Teachers then introduce multiple hands-on investigations, paired with reading experiences with several sources to build on the prior knowledge. Students are continuously journaling to "write about, reflect on, and explain how evidence gathered during authentic science activities links to the concepts being learned," Romance said. Activities culminate with problem-solving and reflection activities. Teachers spend more time on concepts that have broad applicability. Overall, she noted, the curriculum supports cohesion across the science curriculum as students build upon their prior knowledge, consistently add knowledge and depth as they focus on a concept, and use their knowledge about one concept to inform their learning about the next.

Romance shared her experiences with students participating in *Science IDEAS*. In one situation, she found that she had to increase the depth and complexity of experiences for students who had been participating in *Science IDEAS* classrooms in previous years. She also described an experience where students in an at-risk drop-out prevention classroom used a model of the Earth to communicate about why Florida does not experience earthquakes. As she said in her presentation, a member of the press observing this class asked if it was composed of gifted students, to which she replied, "Yes, they are."

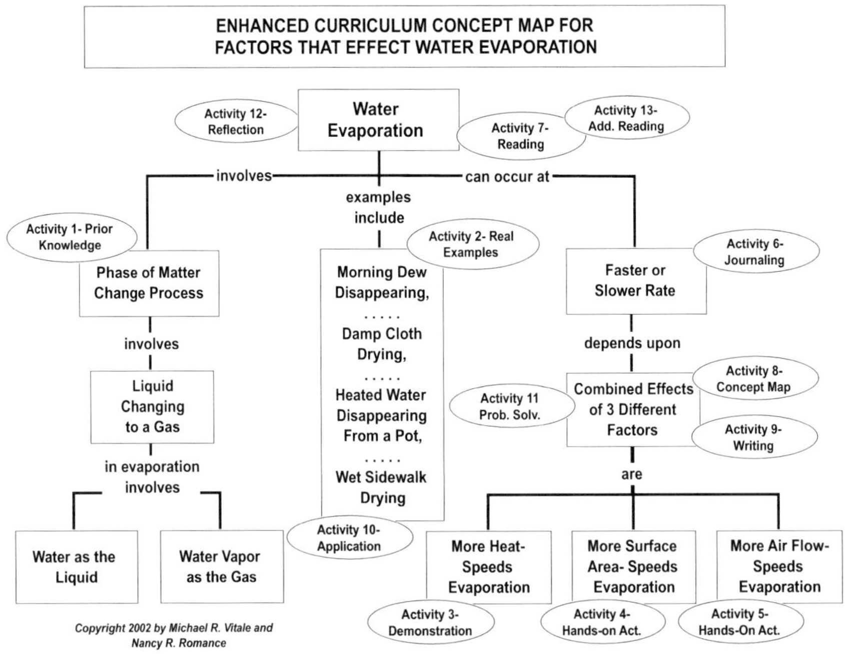

FIGURE 4-1 Example of a propositional concept map for grades 3-5 from *Science IDEAS.* SOURCE: Vitale and Romance (2013).

Two separate longitudinal studies of *Science IDEAS* (2002-2007 and 2003-2008) indicate that students who participate in this curriculum when compared with students in the control group outperform their counterparts in science and reading as measured by the Iowa Test of Basic Skills (Romance and Vitale, 2011). Moreover, these differences are long-lasting and increase over time when measured through the 7th or 8th grades. More limited adaptations of the curriculum targeting 1st- and 2nd-graders also show promising results when compared with control students. As Romance remarked, the results indicate that to improve science outcomes in middle school, efforts must start in elementary school.

MIDDLE SCHOOL EXEMPLAR: INVESTIGATING AND QUESTIONING OUR WORLD THROUGH SCIENCE AND TECHNOLOGY (*IQWST*)

LeeAnn Sutherland, University of Michigan, described *Investigating and Questioning our World through Science and Technology (IQWST),*[3] a middle school curriculum that integrates science and literacy. *IQWST* is a research-based curriculum composed of 12 units across the middle school years. Units in physical science, chemistry, earth science, and life science across three levels focus on answering "driving questions," as shown in Figure 4-2. Each 8- to 10-week unit focused on these questions provides coherence to the development of knowledge along a "storyline" and greater facility with scientific practices. In addition, focus around answering these driving questions encourages student engagement and builds upon students' prior knowledge and experience. Students are actively engaged in making sense of scientific phenomena.

With *IQWST*, students meet the Common Core State Standards for English Language Arts (CCSS for ELA) literacy in science standards through reading, writing, listening, and attending to language in every lesson, according to Sutherland. They also use key scientific practices as specified in the NGSS, including analyzing and interpreting data, developing and using models, constructing explanations, and engaging in argument from evidence. *IQWST* helps students develop proficiency in using the language of science over the course of the middle school years. For example, initially 6th-graders may focus on the nature of evidence and the need for evidence to support their ideas. Over time, students build upon their increasing abilities to develop broader skills in making claims, gathering evidence, and using reasoning skills to evaluate the evidence. These practices are interwoven throughout the curriculum, rather than practiced in isolation, and focus on increasing the depth of student knowledge and skills in scientific practices.

As Sutherland explained, engaging directly with scientific investigation, reading, writing, and talking are components of each *IQWST* lesson. Students also encounter texts through a student IQWST book with embedded questions, procedures, and worksheets. A companion teacher edition assists teachers in introducing and following up with students on their reading. Students then obtain information from their experiences with investigation and text, evaluate the evidence, develop explanations to answer the driving questions, and communicate their

---

[3]For additional information about IQWST, see http://sites.nationalacademies.org/DBASSE/BOSE/DBASSE_085962 [March 2014].

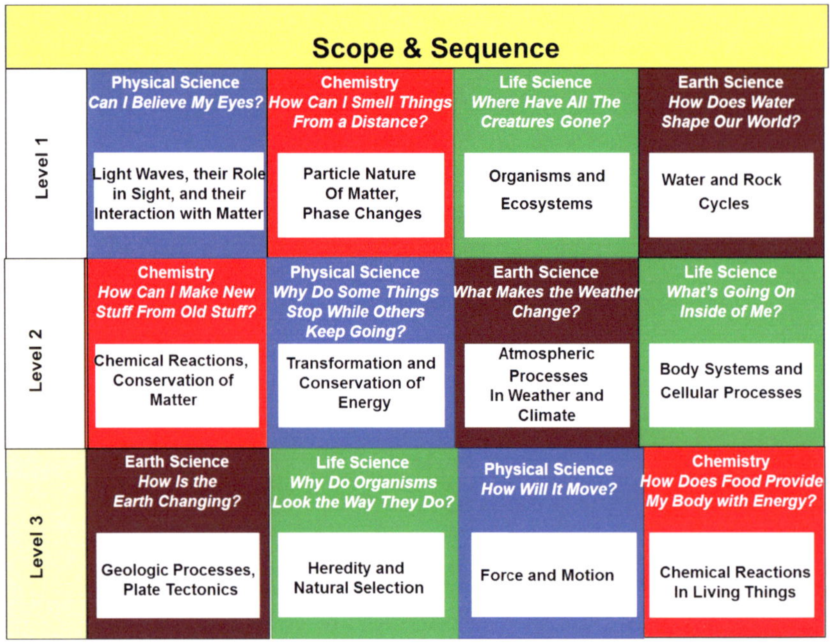

FIGURE 4-2 Scope and sequence of the *IQWST* curriculum.

NOTE: *IQWST, Investigating and Questioning our World through Science and Technology.*

SOURCE: Krajcik et al. (2011).

understandings through talk and writing. Students have opportunities to work individually and collaboratively with other students.

Deborah Peek-Brown, University of Michigan, provided illustrations of how *IQWST* engages students in obtaining evidence, constructing explanations, and engaging in argumentation. At the beginning of a unit, for example, students often engage in a hands-on investigation, build additional conceptual understanding through reading, and then communicate their initial understanding of a phenomenon through writing and talk. *IQWST* reading and writing practices are designed to directly meet CCSS for ELA in these areas.

A significant focus of *IQWST*, according to Peek-Brown, is helping students learn to think and communicate with others using scientific explanations and argumentation. Using claims, evidence, and reasoning to construct explanations and communicate them to others is important not just in science, but also in everyday life. In her words, "What we really want for scientific literacy is for the students to understand that there is a purpose behind this . . . this is the way we think about things."

Peek-Brown explained that teachers model and support scientific discourse to help students learn this language of argumentation. During class discussions, teachers frequently prompt students to provide evidence for their ideas. Over time, students often begin to question each other in a similar manner. Students learn that claims must be supported by appropriate and sufficient data based upon what is already known in science.

Students revisit their earlier explanations and engage in an iterative process of evaluating their own writing and the writing of others using the knowledge that they are gaining through reading and investigation, Peek-Brown stated. Working individually or in pairs, students evaluate whether the written explanation contains adequate evidence and good reasoning. Classroom talk is supported so that students learn to question each other about what supports their line of thinking. Such tasks are designed to emphasize the importance of being able to communicate understanding to others in ways that they can follow. Such tasks address CCSS for ELA in speaking and listening as well as language. *IQWST* involves the same modalities used in learning concepts in its assessments. For example, students may be asked to collect data from a model, draw models, and/or explain in writing their understanding of what is happening.

MIDDLE AND HIGH SCHOOL EXEMPLAR: PROJECT READI

Cynthia Greenleaf of WestEd presented the theory, key features, and examples of *Project READi*, Reading, Evidence, and Argumentation in Disciplinary Instruction.[4] This curriculum for grades 6-12 focuses on building students' ability to read for understanding in science, which she defined as the "capacity to use evidence from multiple sources to construct, justify, and critique models or explanations of science phenomena." It consists of text-based modules that supplement an existing science curriculum, as well as learning progressions, assessment tools, and ongoing professional development.

Greenleaf explained that reading and writing for investigation consists of focusing on evidence and counter-evidence, maintaining a skeptical stance, and attending to details around mechanisms, interactions, and the like. Students are actively engaged with multiple sources and forms of text for the purpose of coming up with explanations that answer questions like, How do we know?, Why do we think differently from one another?, and How can we adjudicate our ideas?

[4]For additional information about Project READi, see http://sites.nationalacademies.org/DBASSE/BOSE/DBASSE_085962 [January 2014].

Of particular importance in *Project READi* is an explicit focus on helping students develop a particular epistemological stance, a way of consciously thinking about how they will approach science reading. According to Greenleaf, this means that students ideally approach a science journal thinking, "I'm going to be confronted with something that might be new or that might put me in the position of having to question existing knowledge, but I'm going to be skeptical about what's there unless and until it is compelling based on evidence."

The development of *Project READi* was motivated by a desire to improve the state of science teaching and specifically to help science teachers use text for more authentic purposes. Greenleaf stated that currently little true science takes place in many secondary science classrooms, where the focus becomes delivering content and teaching *about* science. Texts used often only consist of the textbook, according to her; further, students rarely use scientific argumentation and teachers are not utilizing strategies that enable text-based investigations. Thus, the purpose of *Project READi* was to meet the simultaneous challenge of developing students' science knowledge, engagement, participation in science practices, and ability to read for understanding in science, along with developing science teachers' understanding of science practices, literacy practices, various texts, and repertoire of pedagogical strategies. Developers of this approach want students to learn that "science changes [but] knowledge builds," Greenleaf said.

Science and literacy are intertwined creating a strand of inquiry, as described by Greenleaf. The student learning goals are shown in Box 4-1. Meant to support an existing curriculum with first-hand science experiences, *Project READi* provides students with numerous experiences to read scientific texts, to grapple with the language, and to build a repertoire of sense-making skills. Students engage with scientific texts that have been carefully selected and sequenced, but have not been reduced or simplified in any way.

By way of example, Greenleaf described in some detail a unit in which students investigated the causes of Methicillin-resistant *Staphylococcus aureus* (MRSA). She showed examples of how students are presented with an initial set of questions and readings. Students use the texts to gain knowledge and to create and revisit explanations for the initial questions, as well as generate their own new questions. As students read, teachers help them learn and use active reading strategies, and devote class time to discussing confusing concepts and challenging words. Over the course of the unit, students have multiple opportunities to engage in argumentation about the best explanations. In the case of the MRSA unit, texts are intentionally sequenced so that students encounter elements of the

BOX 4-1

PROJECT READI STUDENT LEARNING GOALS

1—Engage in close reading of a range of science representations; identify, analyze and interpret scientific evidence in texts/sources including graphs, diagrams, models, exposition.

2—Synthesize evidence and information across multiple sources including graphs, diagrams, models, exposition.

3, 4, and 5—Construct, justify, and critique explanations and explanatory models of science phenomena from scientific evidence drawn from multiple courses and using science principles, frameworks, and enduring understandings.

6—Demonstrate understanding of the epistemology of science through inquiry dispositions and conceptual change awareness/orientation, seeking "best understandings giving the evidence."

SOURCE: Adapted from Greenleaf et al. (2013).

causal model for how MRSA is transmitted across the unit. Ultimately, students use their knowledge to construct models and negotiate the best explanations with their classmates. This resolution often results in more questions. Assessment tasks parallel the types of activities students engage in throughout the unit.

As *Project READi* is implemented, Greenleaf said, its developers have identified a number of challenges, as well as benefits. Teachers grapple with balancing to cover the content, while also giving students the time they need to engage with the texts. At the same time, both teachers and students learn from the approach how to conduct true scientific investigations with text, and teachers learn how to turn over sense-making to students. She suggested that adopting these new teaching strategies is challenging for both teachers and students because in many cases it represents a significant departure from their current way of interacting in the classroom and using texts. *Project READi* offers materials to support the curriculum, but Greenleaf shared that teachers need "professional communities and support" to master this approach. A further challenge involves shifting the ways in which models and other scientific representations are created and used, ideally moving away from a focus on aesthetics and toward accurate representation of constructs and usefulness. Helping teachers to learn to use texts by engaging them in opportunities to learn in the same way their students learn has promoted deeper understandings of texts and the practices of science. Greenleaf closed her remarks reinforcing that it is indeed possible to intertwine literacy and scientific practices in authentic ways that address both the CCSS for ELA and NGSS.

DISCUSSION

The workshop brought together these science curricula ranging from early elementary school through high school because they have successfully interwoven science and literacy in meaningful and authentic ways. In the discussions that followed the presentations, presenters responded to a question about how these ideas and approaches could extend to younger children who are still learning to read. Barber and Romance indicated that they are currently working to extend their curricula to this age group, as well as to develop parallel support for teachers at this level. Sutherland and others added that learning to read does not end at the third grade after which students are said to be "reading to learn." Susan Pimentel suggested that read-alouds could be an important way to address limitations in reading among the youngest children, while also building their skills in listening and speaking. Finally, David Pearson cautioned that simplifying the language of science for young children can oversimplify the concepts leading to misinformation. He affirmed that through whatever modality children encounter science texts, it should increase knowledge.

Another discussion focused on how curriculum addressing these scientific practices also addresses the disciplinary core ideas named in the NGSS. Referring to *Project READi*, Susan Goldman indicated that identifying the "big ideas" and core concepts was an integral part of curriculum development. Greenleaf also stated that some of the core constructs included change over time, systems and interactions, and causal mechanisms present in nature. Brian Reiser and Greenleaf also added that certain content lends itself to inclusion in text-based investigations. For example, problems that are particularly engaging to the students, that cannot readily be encountered in a hands-on investigation (e.g., MRSA), and that have abundant readily available data are good candidates for scientific investigations using text.

A final challenge the group discussed in implementing these types of science curricula was the need for change in classroom culture. Students bring their own dispositional factors, Greenleaf noted, such as a sense of self-efficacy or a desire to grapple with difficult problems. Teachers also have a role to play in creating a safe environment, where it is acceptable to make mistakes and to not know the answer, according to Goldman. Further, moving away from recitation to engaging in science involves helping teachers "let go," allowing students to "meander" toward developing the theories. Goldman noted that this takes modeling and scaffolding for teachers. Greenleaf added that engaging teachers in the same process that students go through with text appears to help teachers see the value in allowing students to construct meaning for themselves.

5

PREPARING TEACHERS TO EFFECTIVELY INTERWEAVE SCIENCE AND LITERACY INSTRUCTION

A key challenge for enacting the vision of weaving science and literacy practices together effectively is preparing new and existing teachers. The workshop featured five case studies of programs that are successfully preparing science teachers to have the content knowledge, pedagogical strategies, and orientation to ways of building knowledge put forth in the Next Generation Science Standards (NGSS), while also naturally and effectively weaving in literacy practices. Two of these case studies featured teacher preparation programs in university settings, and the remaining three featured different approaches to professional development for practicing teachers.

Each of the cases offered unique perspectives and approaches; however, several themes emerged from the presentations:

- Working in community and engaging in hands-on experiences along with reading, writing, and speaking are important for learning for both teachers and their students.
- Helping teachers to become aware of the strategies they use for sense-making can help them better understand their students' needs.
- Learning how to construct a supportive classroom culture is an important element for successfully implementing new science practices.

PREPARING NOVICE TEACHERS

Case Study 1: Teacher Preparation at the University of Michigan

Elizabeth Davis, University of Michigan, described the preparation that novice teachers need to successfully weave science and literacy instruction together for their students.[1] According to the NGSS, she stated, students need to have knowledge of core concepts and be able to use the eight science and engineering practices. So, it follows that their teachers need to be proficient in the eight practices, as shown in Box 5-1. However, to be effective, she said teachers must also have content knowledge specifically related to teaching, such as the common difficulties that students encounter and the strengths and weaknesses of various ways of representing science ideas. In addition, teachers also need a repertoire of teaching strategies that map onto NGSS and Common Core State Standards for English Language Arts (CCSS for ELA). In science, teachers need to develop strategies around supporting classroom discourse and eliciting student ideas. In domains emphasized in the CCSS for ELA, teachers also need knowledge of and understandings in literacy, including both the nature of literacy and pedagogical knowledge and strategies. Davis emphasized that these demands indicate the complexity of what novice teachers must master.

Davis described the teacher preparation program at the University of Michigan to illustrate one approach to meeting this challenge. For students at all levels, the teacher preparation program follows a coherent sequence of coursework that involves science methods courses, addresses differentiation of instruction, and, particularly germane to the topic of the workshop, includes a course in literacy in science. For secondary teacher education, the literacy course is not a general course about literacy development and strategies. Rather, it is specifically grounded in using literacy for teaching science. As Davis described, students learn specific techniques for helping children access and comprehend text in science. Further, the teacher preparation program has a practice-based orientation that consists of carefully planned opportunities to practice particular skills in teaching that increase in length and complexity over the course of the program.

Overall, Davis stated that the literacy for science aspect of the University of Michigan secondary teacher preparation program was designed to meet three main goals: (1) to develop recognition that scientific work is infused with literacy practices; (2) to expand students' definition of text to include graphs, diagrams,

[1]For more information, see the following commissioned paper: Davis and Bricker (2013).

BOX 5-1
NGSS SCIENCE AND ENGINEERING PRACTICES

1. Asking questions (for science) and defining problems (for engineering).
2. Developing and using models.
3. Planning and carrying out investigations.
4. Analyzing and interpreting data.
5. Using mathematics and computational thinking.
6. Constructing explanations (for science) and designing solutions (for engineering).
7. Engaging in argument from evidence.
8. Obtaining, evaluating, and communicating information.

NOTE: NGSS, Next Generation Science Standards.

SOURCE: National Research Council (2012).

models, and other representations; and (3) to prepare students to bridge everyday and scientific discourse in the classroom. She emphasized that these goals are necessary because often novice science teachers do not see the value of text and other forms of literacy in the science classroom.

Next, Davis described the University of Michigan novice teacher preparation program for elementary education students. As with the secondary teacher education program, the elementary education program is carefully sequenced. A focus on disciplinary literacy is infused throughout the program. Students begin their teacher preparation program with a course entitled "Children as Sensemakers." This course supports novice teachers in seeing that "children are constantly making sense of the world" and helping them to develop the "knowledge and skills that are related to mediating that sense-making," Davis explained. It also offers them supported opportunities to practice their emerging skills in combining text and hands-on experiences to help children make sense of scientific phenomena.

Throughout the elementary teacher education program, the students build skills toward using informational texts, in preparation to meet the CCSS for ELA, but also in supporting children in obtaining, evaluating, and communicating scientific information. By way of example, Davis indicated that their teacher education students, or interns, learn how to assist students in comparing readings with physical models, as well as how to support students in making sense of their observations using writing in science journals and classroom talk.

In sum, Davis suggested that novice teacher preparation programs must provide students with disciplinary and pedagogical content knowledge, a set of high-leverage teaching strategies, and an understanding of their ethical obligations as teachers. Box 5-2 shows a set of experiences, strategies, and skills that novice teachers need in these programs to enact the visions of the NGSS and CCSS for ELA in Davis' view. Specifically, she indicated that teacher education can use methods such as video exemplars, decompositions of strategies, and opportunities to approximate ideal practice. Preparation programs can also specifically address the "claim-evidence-reasoning" framework and teacher roles in productive classroom discourse. Last, Davis argued that programs can infuse disciplinary literacy throughout their programs by examining literacy-related products, developing discourse conventions, and supporting the infusion of literacy work into science investigations.

BOX 5-2
IMPLICATIONS OF NGSS AND CCSS FOR ELA FOR TEACHER EDUCATION

Novice teachers need to be able to:

- Hear and see the science in students' talk, artifacts, and writing.
- Develop discourse norms that allow students to talk and write science.
- Develop and use scaffolding to support students in science-and literacy practices.
- Use, find, interpret, and evaluate informational text, and support students in doing so to generate, use, and evaluate a *wide range* of texts, including representations of ideas and of data, and support students in understanding these.
- Do all these things to support *all of the students* in the classroom.

NOTES: CCSS for ELA, Common Core State Standards for English Language Arts; NGSS, Next Generation Science Standards.

SOURCE: Davis and Bricker (2013).

Case Study 2: Teacher Preparation at the University of Washington

Mark Windschitl described how the secondary science teacher preparation program at the University of Washington helps novice teachers develop a repertoire of literacy support strategies.[2] A graduate of that program, Lindsey Berk, currently teaching at Chinook Middle School in SeaTac, Washington, provided insight based on her experiences as well. Overall, Windschitl indicated that their teacher education program follows a model of four core teaching strategies, as shown in Figure 5-1, focusing on developing high-leverage strategies. As the model shows, they fit together and follow a sequence.

This approach to preparing science teachers is based ultimately upon what the goals are for students in the classroom. Windschitl indicated that in a science classroom where students are all contributing to knowledge production, students need relevant and compelling contexts for engaging in science, skills in representation that enable them to make their thinking visible to others, scaffolds and routines to facilitate science reading and writing, and sufficient time and opportunity to participate in refining ideas. He specifically addressed how literacy practices are woven in each of the four parts of their preparation model and used in the planning and implementing of 2- to 3-week science units.

Novice teachers first focus on planning for engagement with science ideas, Windschitl explained. A key aspect of this process is selecting a phenomenon that is relevant, compelling, and complex to explain. Topics that are based in phenomena that have great explanatory power are ideal, he said, because figuring out how to make sense of the selected event, observation, or phenomenon serves as the driving force behind all of the lessons in a given unit. The purpose is not to have students reproduce an answer from a textbook, but rather to plan experiences that allow students to generate ideas, gather evidence, develop explanations, and refine their ideas to produce an evidence-based explanation. The teacher preparation program involves helping novices unpack the existing curriculum and the NGSS to identify science ideas around which to plan "anchoring events."

Next, Berk offered an example of an anchoring event that she planned. She focused on the spread of English ivy as an invasive species and trying to determine why it was harmful. Her students made observations and collected samples of English ivy. All subsequent activities, including readings, writing, discussions, and hands-on activities in the unit, were in service of understanding this anchor-

[2]For more information, see the following commissioned paper: Windschitl and Carlson (2013).

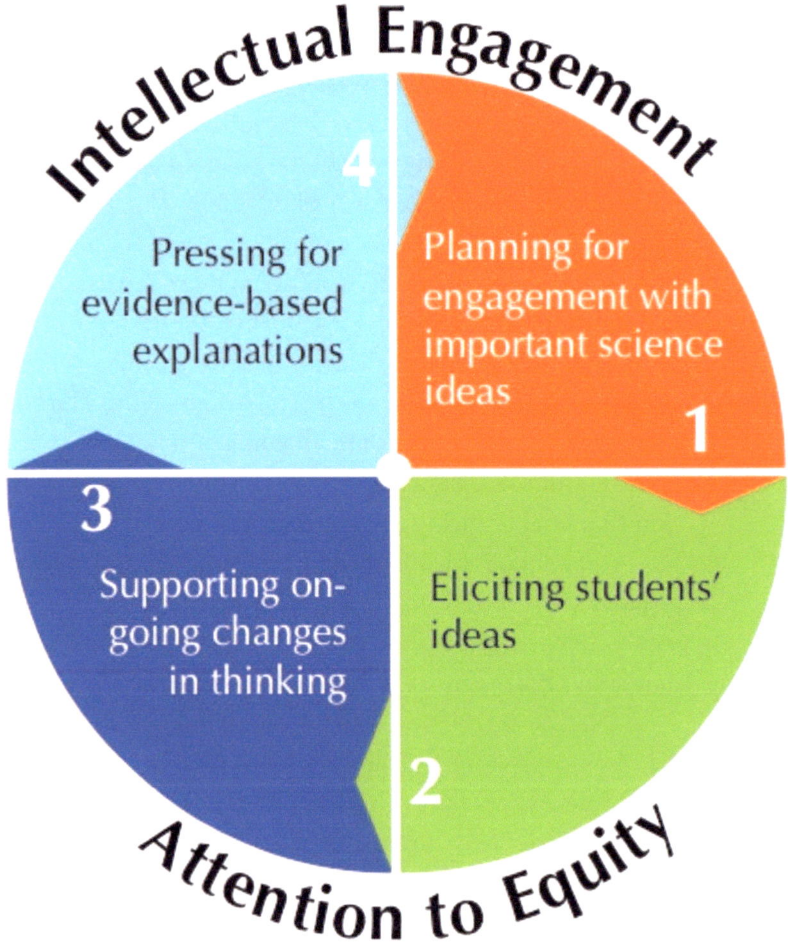

FIGURE 5-1 Model for science teacher preparation at the University of Washington.
SOURCE: Berk and Windschitl (2013).

ing, and culminated in developing a scientific explanation for this particular phenomenon.

After planning for engagement, novice science teachers focus on learning how to elicit student ideas, Berk stated. An important overarching element to this phase is the idea that teaching must specifically attend to ensuring that *all* students are able to contribute and access the curriculum. Eliciting student ideas can begin with telling a story, showing an image, or viewing a video about the anchoring event. The discourse that follows requires that teachers have a set of skills in this domain. Discourse is so central to the approach of this program that students receive a primer on this topic on the first day of their teaching methods course. Further, the program emphasizes "discourse practices" rather than "teaching practices."

Novice teachers learn how to provide students with a structure for representing their ideas and making their thinking available to others. In particular, they guide students in developing pictorial models accompanied by text that explain the phenomenon of the anchoring event. Each student produces a model that shows an event or a process and depicts change over time. They must also label what is observable in their model, as well as what is *not* observable because, as Windschitl stated, "in the world of science, what is unobservable nearly always explains what is observable in the world." These models help students communicate their ideas in ways that practicing scientists use.

Berk illustrated the process of eliciting student ideas by describing a unit she created about force and motion. In this unit, students worked to explain how a person could do a "wall flip," running up a wall to do a back flip. Each student created his or her own pictorial model and text explanations. Students discussed their ideas, and a class poster depicting the initial model was developed. This model became an object for student revision and refinement of ideas as they gained more knowledge over the course of the unit. Berk added that using pencil to create this initial model reinforces the idea that it will be revised over time. Figure 5-2 shows the scaffolds and the ways in which students shared and refined their ideas across the unit, and are described in more detail below.

The third skill that novice teachers learn is supporting students' ongoing changes in thinking. Berk stated that this phase of learning in a unit constitutes the greatest amount of time. Often six or seven different activities can accompany this development of knowledge and understanding. Students may engage in a hands-on activity followed by reading or vice versa. An important part of preparing novice teachers for helping their students engage meaningfully with science texts is helping them become conscious of the processes they use themselves to make sense of texts. Explicitly addressing this awareness helps novice teachers more greatly appreciate what their students who do not yet use these strategies must learn to do. To help students learn these strategies, teachers use scaffolds during prereading, reading, and postreading.

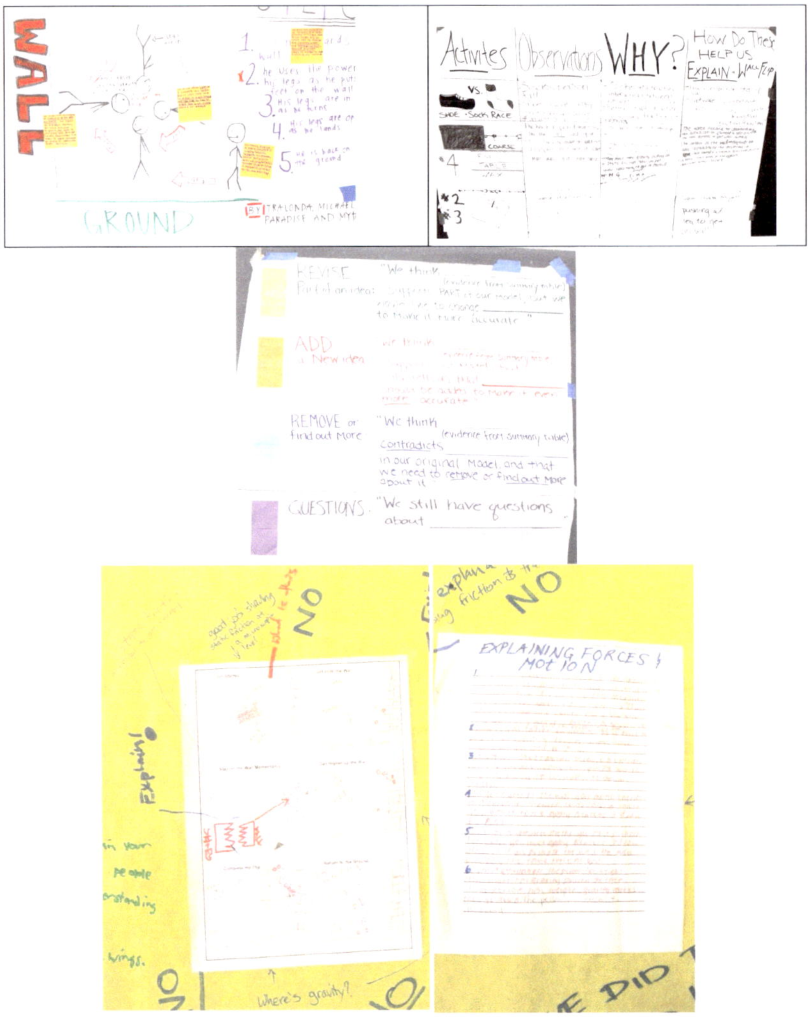

FIGURE 5-2 Examples of scaffolds and student work.
SOURCE: Berk and Windschitl (2013).

Berk showed an example of a scaffold to support changes in thinking during her force and motion unit. She supported students in their creation of a poster that had columns for activities, observations students made during those activities, explanations for their observations based on readings of various texts, and ideas about how that activity contributed to understanding of the anchoring event.

Students worked in groups to compare their individual ideas for each cell of the chart and through discourse, students identified the best information to go on the class poster. Thus, readings are explicitly connected to first-hand investigation, and are always contextualized in service of explaining a compelling scientific phenomenon. Students produce all aspects of the class poster. As Berk stated, "Students get practice at taking complex ideas, making them concise and hearing how other students would put those ideas in one or two sentences."

Berk said the revision process is a key aspect of supporting students' change in thinking; however, most students do not know how to critique their own models or those of other students in the way that scientists do. Moreover, the models can be very complex, and talk around sense-making can move too quickly for many students, particularly for English language learners. To address these issues around the revision process, Berk developed a model while studying at the University of Washington that has become widely adopted in her area. She developed a system of using color-coded sticky notes, accompanied by a set of sentence stems to scaffold support for revisions. Four different colors were used to represent "revise part of an idea," "add a new idea," "remove or find out more," and "questions." Students wrote their ideas on the sticky notes using scientific language and terms and then placed them on the part of the model on the poster needing the revision. This method also ensures that the pace slows down to provide sufficient time for thinking and composing feedback to peers. She added that her students conduct one model revision to avoid "model fatigue" that can result from pressing students to revise two or more times.

The last three days of a unit are devoted to pressing for evidence-based explanations of the anchoring event. The purpose is to help students pull together the multiple models, texts, and revisions to create a final model. Writing these final causal explanations requires scaffolding, according to Windschitl. Necessary supports include providing a structure for writing these explanations, sufficient time to do so, prewriting activities, opportunities to rehearse their explanations, and strategies for breaking explanations into their smaller components. A key question that helps guide students as they begin to craft explanations is, "What is it you think we are trying to explain?" One strategy that Berk used to help students rehearse their final explanations was having students create new, more intricate pictorial models and text. She then had students post their papers on a larger poster, on which their peers could write feedback and ideas for further refining their final explanations.

PROFESSIONAL DEVELOPMENT WITH PRACTICING TEACHERS

Case Study 3: Next Generation Science Exemplar-Based Professional Learning Systems

Jean Moon, Tidemark Institute, described a professional development model that she developed with a team of researchers.[3] The model, Next Generation Science Exemplar-Based Professional Learning Systems (NGSX), is housed on a Web-based platform and based on video exemplars.[4] Several principles guided the design of NGSX, Moon said. First, the model focuses on science. Participating teachers learn disciplinary core ideas in science and engage in scientific practices, such as explanation, argument, and modeling. Second, NGSX emphasizes student sense-making. Teachers analyze student discourse and work using video cases as they build core ideas and strategies. Third, NGSX explicitly addresses pedagogy and how teachers can support student practices and discourse, again using video exemplars. In Moon's words, an approach that includes video exemplars helps to "get at something that's very critical in getting to this new vision and that's helping teachers imagine what this looks like and what it feels like."

NGSX is a blended learning model organized into learning pathways, according to Moon. An array of resources, experts, tools, and tasks are all housed on a Web platform. Groups of teachers, who teach at all levels from elementary through high school, meet face-to-face in groups facilitated by one of the teachers in the group. Teachers then access the Web-based materials via laptop, smart phone, or tablet. Groups establish their own pace through the pathways, which each consist of eight or nine units. Each unit takes approximately three to four hours with additional "on your own" activities that tie previous units with current and future ones.

Moon described how teachers begin each pathway. They start by viewing a video that presents them with a challenge about a particular science construct that they will learn about. Teachers work together to progress through the pathway, but continue interface with the Website, uploading and posting pictures of their work. They also encounter expertise from scientists, as well as from pedagogical experts. The Website also possesses tools that help to "catalyze" social interaction.

For the initial pilot of NGSX, Units 1 to 3 focused directly on science content and centered on the following question: "What are models in science, and

[3]For more information, see the following commissioned paper: Moon (2013).

[4]For more information, visit http://www.ngsx.org/ [March 2014].

how are they evaluated and revisited?" Next, Units 4 and 5 addressed two questions: "How do I build classroom culture that supports public reasoning?" and "How do I build a classroom culture that supports all learners?" These two units emphasized the culture of scientific discourse in the classroom. Units 6 and 7 focused on argumentation and how to help students argue from evidence, as well as the types of tools teachers can use to help students refine models over time to develop deep explanations of phenomena. In each unit, Moon emphasized that science and literacy are "all very integrated." She also stated that NGSX aims to situate professional development for teachers that is contextualized as closely as possible to teachers' own classrooms and students. Throughout the professional development, teachers do experiments, talk with one another, write, and refine their ideas, documenting their experiences with photos and videos. Moon also indicated that she and her colleagues are using NGSX as a context for research.

Jocelyn Lloyd, a 1st-grade teacher at Woodland Academy in Worcester, Massachusetts, described her experiences as a recent participant in NGSX. Specific aspects of the professional development were particularly positive in her view. She noted that she and her colleagues were engaged in hands-on experiences right from the beginning, which contributed to her feeling that she "forgot she was in professional development." The approach also made use of strategies that Lloyd has found useful with her students, namely learning by doing, but also processing information with others. Lloyd also appreciated having colleagues as facilitators, which she said fostered team-building. She said it also eliminated feelings of intimidation that can occur with an expert leader. With this group dynamic, she felt that it was acceptable not to have all the answers.

Lloyd then described in more detail the particular science content she encountered through NGSX, states of matter, and how she applied her new knowledge and strategies in her own classroom. She added that her classroom is composed of 22 students, 15 of whom are English language learners (ELLs) and all of whom receive free and reduced-cost school lunches. Her goals for her students not only focused on helping to learn about states of matter, but also helping the students learn to discuss, debate, predict, and collaborate to make sense of their observations. Lloyd paired several hands-on experiences with productive classroom discourse to share ideas. She indicated that she was able to use comparable types of experiences in writing in her classroom as she experienced herself in NGSX. In closing, she emphasized that the hands-on approach to professional development assisted her in internalizing the approaches to teaching science and in bringing the practices back to her classroom.

Case Study 4: Quality Teaching of English Learners at the International Newcomer Academy

Aida Walqui, WestEd, presented a professional development model for enacting literacy for science that she has used over the last three years at the International Newcomer Academy (INA) in Fort Worth, Texas, a school of more than 300 students who have just arrived in the United States from more than 35 different countries. More than half of the students are Spanish-speaking, but 25 other languages are spoken as well. She began her remarks indicating that the experiences of the partnership between Quality Teaching for English Learners (Q-TEL) and INA shows how professional development can support ambitious science learning with ELLs, who are present to varying degrees in all schools throughout the United States. She stated, "We used to think that kids needed English first, but now we see that students engaged with worthwhile science practices . . . also develop the ways linguistically of enacting those practices and developing literacy skills."

Walqui indicated that Q-TEL is based on several key premises about English language learners (ELLs) that guide their work. First, they come to school with great potential, and it is critical to avoid deficit-based views of these students. Building on this idea, she stated, "Our role as educators is to grow [a student's] potential through apprenticeship processes that work beyond their level of autonomy and both challenge and support students in their gradual appropriation of practices." This mindset frames their "pedagogy of promise," where students' future success is assumed. To achieve this vision for English language learners, Q-TEL centers its approach on providing scaffolding that supports students and teachers. Walqui argued that teachers who possess the knowledge and strategies to serve the needs of ELL students to learn science well will serve all students well. However, the opposite is not the case, she stated.

Teachers who are to enact this vision of ambitious science learning for all need support that provides them with multiple opportunities to deepen their subject matter knowledge, grow their expertise in disciplinary pedagogy, learn to use existing curricula critically, tailor instruction responsively to context and student needs, and participate in learning communities of other teachers, according to Walqui. Even teachers who have participated in excellent teacher preparation programs continue to need these opportunities, she argued. With regard to literacy for science in particular, teachers need to continue to deepen their knowledge as well as their awareness of literacy practices as they are enacted when they read and use language. This thoughtfulness requires devoting time for planning and implementing. Overall, Walqui stressed that professional development that succeeds in building teacher knowledge, expertise, and strategies cannot be achieved through a workshop, but rather requires time, ongoing support, and an orientation toward lifelong learning.

Walqui described the key features of the professional development program at INA. The school-wide program is long term and intensive over a three-year period. All teachers at INA participate in learning communities situated within their particular disciplines, and they experience a coherent set of supports that range from workshops to intensive coaching. The content of the professional development is multilayered and theory-driven, focusing both on disciplinary content and quality classroom interactions. Central to the Q-TEL approach is support for student "apprentices" as they move toward greater conceptual knowledge and facility with practices. Walqui emphasized that students must be actively invited to participate and supported with scaffolds that fade as independence increases.

Professional development at INA is nested, as shown in Figure 5-3, and designed to build institutional capacity. Walqui detailed the process. Facilitators spend six days working with educational leaders at the school, followed by one day with all staff across disciplines and five days with each team of teachers within a discipline. During these whole-day training sessions that teachers spend with their discipline-specific teams, they work on exemplar lessons learning scaffolding techniques and a repertoire of interactive strategies for increasing student engagement in the learning process. These exemplar lessons use a balance of hand-on activities, readings, and reflective discussions. Teachers who become coaches for other teachers within their discipline receive an additional four days of training, and some of them go on to become professional developers, spending an additional eight days in training. Coaches work with teachers following the workshops in four-day cycles. They meet with teachers prior to conducting classroom observations, observe, and then meet afterward to debrief. This reflective process enables teachers to see how changes in their strategies lead to changes in student participation and an increase in the rigor of the content. This nested approach builds the in-house capabilities and expertise of the entire school over a three-year period.

Tanya Warren, who has participated in the professional development over the past three years, described her work as a teacher of integrated physics and chemistry at International Newcomer Academy. She indicated that her participation in this professional development has led to dramatic shifts in both her thinking and teaching. For example, she said she previously felt students had to have a

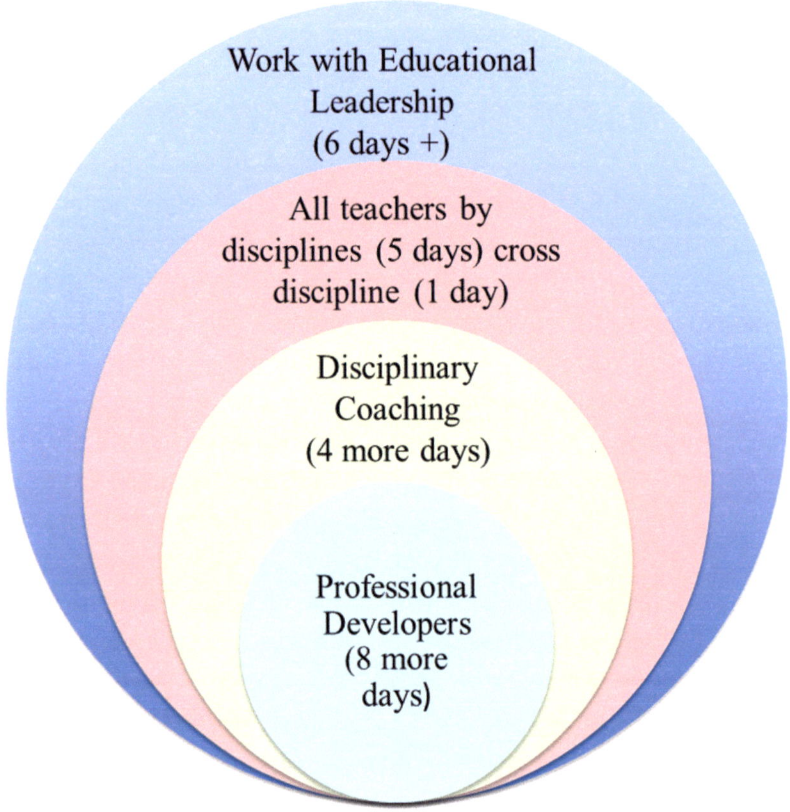

FIGURE 5-3 Nested model of professional development at International Newcomer Academy.
SOURCE: Adapted from Walqui et al. (2013).

certain level of English proficiency to participate in various aspects of the curriculum. Now, she believes that the appropriate level of support is what is needed for all students to engage meaningfully with the material. In her words, "ELLs can be successful at many things I once thought were impossible."

Describing a particular unit of study with her students focused on atomic structure, she shared what her previous approach to instruction would have been. Prior to her professional development, she would have had a teacher-centered approach with a PowerPoint presentation, readings, and a "foldable" writing assignment. Now, even though she continues to use the same curriculum, she has learned to adapt it by putting supports into place so that students directly engage with one other as they build knowledge. She has put into place a structure for various activities within consistent and predictable routines. For example, students make observations and then write questions based on their observations. They then work in small groups to share their questions in a round-robin fashion, with the group working together to reach consensus about one of those questions to share with the whole class. Rules, such as no interrupting, are established to support these routines. The approaches to activities are designed to mimic those of practicing scientists, such as asking questions, making observations, interpreting diagrams, and collaborating with peers. Warren showed videos[5] of students who had recently arrived in the United States engaging in scientific discussions with peers as they worked to refine and clarify their ideas. She emphasized that through this type of discourse, students developed their own voice and increased in confidence as they learned to use the language of science. She noted that the use of two different languages in the discussion did not impede their progress in beginning to enact scientific practices and develop conceptual knowledge. Warren added that her students demonstrated increasing mental stamina as they persevered during the sense-making process, and they learned to listen to one another to construct meaning together. Similarly, she said the professional community of teachers learned to collaborate to reflect and make sense of their own professional practices as science teachers.

Case Study 5: Partnership for Effective Science Teaching and Learning

Brett Moulding, Utah Partnership for Effective Science Teaching and Learning (PESTL), described PESTL's intensive three-year, 330-hour professional development program for science teachers.[6] Its developers used the expectations laid out in *Taking Science to School*, published in 2007, in designing the program, as well as *Ready, Set, SCIENCE!* in their first year of implementation. Since 2008, they have worked with two cohorts of 120 educators (2008-2011 and 2011-2014). Moulding focused his remarks on how they have approached helping teachers develop an understanding of science practices, crosscutting concepts, and core ideas in service of helping students communicate in science. A defining feature of the approach is the focus on "science performances," which are multifaceted, authentic experiences in doing, thinking, and communicating science. He likened

[5]To watch the video, go to time 5:42 of Warren's presentation at http://sites.nationalacademies.org/DBASSE/BOSE/DBASSE_087376 [August 2014].

[6]For more information, see the following commissioned paper: Moulding (2013).

science instruction without performance to watching a piano teacher and reading music but never playing the piano.

As Moulding described, the PESTL approach to professional development engages teachers in doing science performances themselves, as well as reflecting upon their instructional strategies. Overall, Moulding argued that professional development for science teachers must parallel very closely what science students should be doing. Performance in science, supported by instruction, assessment, materials, and professional development, is at the intersection of the disciplinary core ideas, crosscutting concepts, and science and engineering practices.

The professional development provides teachers with a structure for the crosscutting concepts in science. This structure links concepts around causality, patterns, and systems. Focusing on causality helps to support the reading and writing of science and constructing explanations, Moulding noted. Both teachers and students learn to look for patterns throughout their engagement with phenomena, which assists them with constructing and rehearsing their scientific explanations. Systems help in categorizing types of phenomena, including change and stability, matter and energy, and scale and proportion. The combination of defining systems, finding and using patterns for evidence, and determining cause and effect relationships in service of constructing explanations of scientific phenomena has proven to be a powerful tool, according to Moulding.

Moulding then described how this approach to professional development in science addresses the science and engineering practices. Teachers and their students move from gathering evidence to reasoning to communicating. He noted that while reasoning cuts across all areas, their professional development addresses it explicitly in detail through discussion with teachers because often teachers move from gathering to communicating, and "jump over reasoning." These practices directly link to the CCSS for ELA and NGSS, as shown in Table 5-1. Showing these linkages to teachers has helped them understand the nature of evidence across disciplines, he added.

Moulding provided an example of this approach using the phenomenon of the variable timing of when quaking aspen[7] leaves emerge. Teachers engage with this specific example, conducting science performances to explain this phenomenon. The process of constructing an explanation consists of five components:

[7]The quaking aspen is a tree native to North America, characterized by smooth white bark and heart-shaped leaves that "quake" when the wind blows. They grow clustered together in colonies connected by a common root system. Some aspen trees leaf out several weeks before other nearby aspen trees.

TABLE 5-1 Connections among Science, Engineering, and Literacy

Practices	Science and Engineering Practices	Literacy Expectations
Gather	• *Obtain Information* • *Ask Questions/Define Problems* • *Plan and Carry Out Investigations* • *Use Models to Gather Data* • *Use Mathematics and Computational Thinking*	• Ask questions to gain understanding • Obtain information through careful reading of relevant and reliable text, graphs, charts and listening to reliable sources. • Develop and organize ideas, concepts, and observations (data and measurements from investigations).
Reason	• *Evaluate Information* • *Analyze Data* • *Use Mathematics and Computational Thinking* • *Construct Explanations/Solve Problems* • *Develop Arguments from Evidence* • *Use Models to Predict and Develop Evidence*	• *Evaluate information for evidence and relate explanations and arguments to appropriate evidence.* • *Explain how an author uses reasons and evidence to support particular points in a text, identifying which reasons and evidence support which point[s].*
Communicate	• *Communicate Information* • *Use Argument from Evidence (written/oral)* • *Use Models to Communicate*	• Communicate in meaningful ways through speaking and writing that use evidence to support arguments. • Write informative/explanatory texts to examine a topic and convey ideas and information clearly. • Present information, findings, and supporting evidence such that listeners can follow the line of reasoning and the organization, development, and style are appropriate to task, purpose, and audience.

SOURCE: Moulding (2013).

group performance, individual performance, classroom discourse, science reflection, and teacher reflection. The group performance begins with exploring the initial information about the quaking aspen and developing a set of questions. Teachers gather information from a variety of sources and investigate possible explanations for the variable timing of leafing. They develop pieces of evidence, using core ideas to support their emerging explanations.

During individual performance, teachers write their explanations about causality in a science journal. They must include sources of information and the evidence in their explanations. This period of individual work is followed by classroom discourse about the phenomenon. According to Moulding, "We believe that their use of core ideas in science as evidence when connected to explanations is the most powerful shift in that vision for science education brought about. It also becomes a powerful shift in the way students construct writing by putting core ideas into the writing pieces that they do."

Teachers are provided with devices to support obtaining information from valuable sources. Asking questions that do not have simple answers helps to move how teachers think about and obtain information about science ideas, Moulding said. They move from simply looking for information about scientific terms, like hydrogen bonding or evaporation, down to core ideas, like matter and energy. In Moulding's experience, this shift has proven important in helping teachers see that the "world of science is simpler than what we have let them on to believe."

In showing examples from two teachers, Moulding emphasized that their activities explicitly support looking for core concepts and reasoning in students' reading, writing, and speaking. In addition, effective scientific classroom discourse also takes place within a supportive classroom culture with rules and an orientation towards collaboration and cooperation. He also indicated that teachers' written reflections on their strategies suggest that science journaling and having long segments of time devoted to writing about core ideas, evidence, and explanations for phenomenon is a meaningful way for students to engage in the sense-making process.

For science writing to be a compelling activity, both teachers and students need interesting phenomena to write about, Moulding stated. Further, he added, both teachers and students need tools to support writing. A deep understanding of core ideas developed over time, as well as their own direct experiences with phenomena, information gathered through reading, and use of online sources, all support science writing. In addition, structures are put in place to support the use of scientific evidence in writing.

Moulding differentiated the roles of teachers. He noted that the science and engineering practices described in the NGSS constitute what students are to do. However, instructional strategies are what teachers use to develop these practices, tailored to specific scientific performances. He closed by reinforcing that the goal for teachers is for their students to be able to perform science, gather information, reason, and communicate effectively.

DISCUSSION

Following the presentations of the teacher preparation and professional development programs, the audience and presenters engaged in a discussion of crosscutting issues and lingering concerns. During the discussion, an audience member raised the issue of who retains authority for determining knowledge in the classroom. Allowing students to build their knowledge and understanding by gathering evidence, using reasoning, and constructing explanations means that teachers let go of being the source of authority. When teachers retain ultimate authority of knowledge, classroom discourse can focus on students trying to guess what answer the teacher wants, rather than using scientific reasoning and argument to construct the best explanations. Berk responded that students are used to having teachers decide what answers are right and wrong, and that it takes time to redirect class discussions toward the evidence. However, she added, doing so is possible and that when this is achieved, the knowledge of the entire class is elevated.

Additional discussion centered on the time needed for instructional strategies to change and solidify. One participant observed that teachers can "panic," abandoning their planning, particularly in the moment during scientific discourse in the classroom. Lloyd indicated that this is especially difficult when a discussion is moving in an incorrect direction with students agreeing. When teachers face ambiguity or discomfort in these situations, it is easy to revert to previous instruction methods or ways in which teachers themselves were taught. Davis indicated that creating and showing video to teachers of themselves is one way to help teachers reflect upon their instructional strategies and to begin to become aware of when they are shifting away from productive scientific discourse. Warren affirmed this idea, and added that teachers not only need to debrief about their strategies, but also need time, often several years, to solidify their strategies in science teaching.

6

SUPPORTING LITERACY FOR SCIENCE ON VARIOUS SCALES

Building systems that support literacy for science not only within classrooms and schools, but also across schools, communities, districts, and states requires unique strategies and approaches. Several presentations at the workshop addressed issues in working not only with science teachers to effect change, but also with teachers across disciplines, principals, and other administrators. They offered specific strategies used, challenges faced, and lessons learned to help inform future efforts to build knowledge, capacity, and implementation of excellent science practices on a widening scale. The presenters addressed how they are working to build a shared vision to meet the needs of all students.

Across the presentations, the following themes emerged:

- Teachers learning new approaches and strategies were engaged in communities of learning that emulated approaches to be used with students.
- Creating a shared vision for science education, and more specifically literacy for science, required time and engagement with individuals at all levels of the system.
- At the school district and state levels, changing practices was facilitated by supportive policies.
- Efforts to build capacity to support and sustain the changes include training trainers, working with principals, or supporting "trailblazing schools," among other strategies.
- Phased approaches are bringing about change in awareness, vision, and implementation of new practices in science education.

- Efforts are under way to help states address the needs of English language learners (ELLs) as they adopt and implement the Common Core State Standards for English Language Arts (CCSS for ELA) and Common Core State Standards for Mathematics and the Next Generation Science Standards (NGSS).

THE NEW VISIONS FOR PUBLIC SCHOOLS EXAMPLE

Kiran Purohit described New Visions for Public Schools' work with New Visions Charter High School for Advanced Math and Science (AMS), and their lessons learned in enacting a literacy support model specifically for this school.[1] New Visions for Public Schools supports 75 public middle and high schools and 6 charter schools across New York City. The public and charter schools with which they work tend to be demographically similar, with many students performing below grade level, nearly 20 percent with special needs, and approximately 10 percent as ELLs. A new charter school, AMS, began in 2011. In the first year of its existence, students focused on biology in ninth grade, physics in tenth grade, and now chemistry in eleventh grade.

As Purohit explained, the New Visions network has an initiative to support reading and writing across disciplines for the schools it supports, called the Literacy Design Collaborative (LDC).[2] This initiative predates the NGSS, but it is aligned with the CCSS for ELA. The LDC is based on cross-disciplinary research practices and includes a set of template tasks that involve reading, listening and speaking, and writing. One key element of this approach is the "writing cascade." As part of the writing cascade, students complete one major piece of writing of a single type per three-week period in each of several subjects. Purohit showed an example of a writing cascade (see Figure 6-1), which she said helps to ensure that teachers of different subjects are talking about, thinking about, and structuring writing in similar ways, and also that students need only focus on one major piece of writing at a time. Purohit noted that this comparability and staggered approach is especially important for students who struggle in school.

As part of the LDC initiative, teachers work in cross-disciplinary investigation groups, Purohit explained. Following the design of the skill ladders, teachers implement the tasks in their classrooms. The teacher groups then meet to collabo-

[1]Additional information about New Visions' work with AMS is available at http://sites.nationalacademies.org/DBASSE/BOSE/DBASSE_085962 [March 2014].

[2]For more information, see http://www.ldc.org/ [July 2014].

LDC - Example Writing Cascade

	Social Studies	ELA	Science	Math
Week 2- 4	**Informational** *Interview with a religious leader*			
Week 4- 6		**Informational** *What can you work to change?*		
Week 6- 8			**Informational** *Is there life on Mars?*	
Week 8- 10				**Narrative** **Procedural** *Cell Phone Task*

FIGURE 6-1 Example of a writing cascade.
SOURCE: Miller and Purohit (2013).

rate around student work. They use this process to rearticulate the skills on which students are still working. This process functions in a continuous feedback loop.

By engaging in these groups in year 1 of AMS, Purohit noted several challenges that teachers encountered. First, teachers found examining student work across disciplines more challenging than originally anticipated. Second, they found that their approach was still not meeting the needs of the lowest third of their students. Therefore, in year 2, New Visions engaged with AMS in action research to address these concerns to determine how to make science literacy accessible, how to better support student engagement, and how to engage the community in science classrooms effectively.

The action research involved refining the teaching tasks to incorporate the NGSS practices. According to Purohit, the writing tasks were adapted so that students engaged in more frequent writing in "mini-tasks," rather than less frequent writing of major pieces to foster more opportunities for practice. Last, the team focused on building teacher capacity through collaborative planning with experts from the community. Instead of inviting experts to the classroom as guest speakers to supplement instruction, experts assisted teachers with designing lessons and planning tasks at the outset to foster greater authenticity in the way that students engaged with scientific phenomena. This included holding a retreat where teachers and experts examined both content and science practices in the initial design of student tasks, and then engaging with the experts in the revision of tasks.

To illustrate this approach, Danielle Miller of AMS described a unit on "Pests in the City" implemented at AMS. By engaging with experts in entomology, pest control, and housing, the team identified key content, such as life cycles and ecosystems, and practices, such as developing and using models that the unit would address. This approach was also informed by the work of Richard Elmore[3] that emphasizes a three-pronged approach to improving instruction—increasing teacher knowledge and skills, increasing active student engagement, and raising the level of content.

Miller explained that, now that AMS is in its third year of existence and partnership with New Visions, the action research is focusing on student levels of motivation and the link with engagement. In addition, the team has opted to focus on a small subset of science practices at any given point in time during the implementation of the LDC. Finally, they are continuing to further integrate community experts into their classrooms. These innovations are requiring that AMS find new ways to structure and increase planning and meeting time. Aligning these new ways of engaging in science with current high-stakes testing continues to pose a challenge.

THE OAKLAND UNIFIED SCHOOL DISTRICT EXAMPLE

María Santos, Oakland Unified School District (OUSD), presented her experiences in supporting literacy for science on a school-district level in Oakland, California.[4] OUSD has approximately 37,000 students, 80 percent of whom are low-income and one-third of whom are ELLs. The district has established a model for education that emphasizes three key areas: (1) ensuring a high-quality instructional core, (2) creating equitable opportunities for learning, and (3) developing social-emotional and physical health. This system-wide plan involves meeting CCSS for ELA and mathematics, NGSS, and other science, technology, engineering, and mathematics (STEM) initiatives, she said.

[3]For details, see http://www.acsa.org/MainMenuCategories/ProfessionalLearning/TrainingsandEvents/Creating-Academic-Optimism/Session-1/IstructionalCore.aspx [March 2014].

[4]Additional information about the Oakland Unified School District approach to literacy for science is available at http://sites.nationalacademies.org/DBASSE/BOSE/DBASSE_085962 [March 2014].

Santos said having a set of policies that support this model has been essential in moving classroom strategies forward, particularly in ensuring that daily and weekly time specifically devoted to science at the elementary level are standardized. Stable funding streams, opportunities for professional development, technology, and strategic partnerships have also supported efforts to improve science education across OUSD. Other policies that make explicit the support for STEM and that increase the number of science classes that are required for graduation also make clear the priorities of the district.

To enact this vision, Santos continued, district leaders developed a plan for moving from awareness to implementation in a careful and coherent manner, working with individuals at all levels of the system. Phasing in new policies and practices with significant time for discussions with various stakeholders to build common understanding of the vision was essential in Santos' view. She added that the theory of action focused on all of the layers of individuals from principal supervisors, principals, teacher leaders, teachers, and other specialists to emphasize that change is the responsibility of everyone in the system. As OUSD has implemented opportunities for adult learning around implementation of the NGSS, Santos said they have worked to develop environments that are safe for exploring and building knowledge, and where teachers are encouraged to question, push one another, and construct meaning, much as they would like students to do in the classroom. Importantly, principals also engage in education that enables them to know what teaching and learning science as envisioned in the NGSS looks like.

Leaders in OUSD identified three practices that cut across both CCSS for ELA and NGSS and that aligned with their vision for thriving students: (1) close reading of complex text (informational and literary); (2) academic discussion; and (3) evidence-based writing. Santos stated that the selection of just a few practices enabled them to focus on depth of understanding. Moreover, choosing three practices felt manageable to most teachers and principals. Work helping teachers and others develop their skills in fostering academic discussions began in fall 2013.

Santos related that efforts to build classroom cultures supportive of academic discussions were consistent with OUSD's emphasis on social and emotional health. This involved working with teachers and principals to unpack the elements of a safe academic environment where students listen to and respect one another, are able to manage their own time talking and collaborating, and feel free to take risks. Such a supportive culture also builds in time for thinking. Within this culture, teachers look to help students learn how to articulate their reasoning, argue from evidence, use general academic and disciplinary language, and learn to build on, challenge, and revise their ideas and those of others. Always the focus of these system-wide shifts is how to promote children's sense-making abilities, according to Santos.

Development opportunities across the district help teachers and others to access new tools to foster shifts in their classroom practice. One tool Santos described involves equipping teachers with new structures to support academic discussions, such as "think-pair-share" opportunities or discussion circles. Content of the discourse focuses on using evidence, argumentation, and constructing meaning. However, most students require scaffolded support to engage in this type of discourse. So, teachers provide scaffolds to help students elaborate and clarify ideas, strengthen arguments with examples and evidence, and build on and challenge ideas respectfully. Santos noted that without new sets of tools, teachers tend to revert to previous ways of discussing science with their students. She added that new tasks are needed that lend themselves to these richer discussions. An initial step in providing teachers with new tools and tasks that OUSD undertook involved bringing 167 teachers together over the summer to develop new units that afford more opportunities for rich discourse and productive struggle than previous science units.

Santos also shared OUSD's experiences in implementing the use of science notebooks as a tool for sense-making across the district, starting with elementary students. Students use writing and drawing as part of the process of making sense of experiences. They also use the notebooks to record, analyze, and interpret data on their own and with others. Teachers provide students with sentence frames, word banks, and visual scaffolds to assist them in their use of the science notebooks. Teachers also use these notebooks to better understand children's thinking.

Principals and their supervisors also need new tools and strategies to effectively support teachers as they shift their instructional strategies, Santos said. Principals need to fully understand the practices and the common problems that can arise across content areas. They need the data, tools, and resources with which to evaluate their teachers' use of science practices. Further, they must collaborate with others within and across schools. For teachers to change and feel comfortable in taking risks, they need to know that their principals understand and support what they are doing, according to Santos. The use of video and observations during instructional rounds has been important to these efforts. Principals are encouraged to be reflective and plan for next steps for their school. Recently, similar efforts have begun with groups of teacher leaders.

Santos described one tool that principals use in OUSD that has proved useful. Principals use a 5- × 8-inch card focused on the NGSS practices during their observations. Rather than noting teacher behaviors during their observations, principals focus on student behaviors or "vital student actions." When they analyze classrooms, they first look to see what students were doing and then to what classroom conditions and teacher actions supported those student behaviors. Principals also have a guidebook with clear expectations in assessment, curriculum, and instruction, with most attention on instruction. Overall, there is a shift away from focusing on assessments and benchmark data toward thinking about student and instructional strategies. This focus on practices "is major," according to Santos, "because we get a lot of pushback from folks—'Give me the assessment. How am I going to know?'"

Santos concluded her presentation by describing a set of strategies that OUSD uses to advance science practices. First, she described 13 "trailblazer" science and literacy cohort schools that are serving as lab sites for the district. These schools help to build capacity across the district by creating new resources and supports that other schools will be able to use. These schools receive a significant amount of support from the district to enable this work. Second, OUSD holds five-week Summer Institutes. Teachers at the elementary level who attend focus on curriculum, instruction, planning, and pedagogy. One element of the Summer Institute is specifically geared to principals. Third, she described a science writing task specifically designed for 4th- and 5th-grade students. Students engage in this task over a week-long period, ultimately producing a science opinion essay.

THE SOUTH DAKOTA STATEWIDE EXAMPLE

The workshop also featured an example from a state-wide effort to move science practices in K-12 education forward. Sam Shaw, South Dakota Department of Education, provided an overview of his state's approach to data, lessons learned, and next steps toward meeting their goals.[5] He described South Dakota as a state with a small population, including just over 120,000 public school students and about 9,500 public school teachers, spread over a relatively large geographic area. Middle and high school science teachers constitute 875 of the public school teachers in the state.

[5]Additional information about South Dakota's approach to science education is available at http://sites.nationalacademies.org/DBASSE/BOSE/DBASSE_085962 [March 2014].

Shaw then painted a broad vision of his state's progress toward implementing a new vision for science education in South Dakota. Overall, he noted that while progress is moving with a clear vision from *A Framework for K-12 Science Education* (K-12 framework, National Research Council, 2012), the state's population of science teachers as a whole are further behind in their vision and approaches. Teachers are more likely to continue to be informed by the *National Science Education Standards* (National Research Council, 1996). Because of this disparity of vision, Shaw has focused his efforts as a state leader on "bringing people up to speed" and creating a shared vision for science education across the state based on the K-12 framework.

Part 1 of this initiative was the development of science academies aided in part by the governor's "Investing in Teachers Package." This was driven largely by adoption of CCSS, but widened in scope to address needs in science at the governor's request. Shaw worked with the governor to emphasize student performance, emphasizing a conceptual shift away from teaching science facts. At the science academies, 22 teachers were trained as facilitators in 2012 over a week-long period. During that training, these teachers learned about the K-12 framework; yet, follow-up with the trainers through video indicated that wide variation among the teachers regarding the vision for science education still existed. Although this was initially challenging, their timeline for implementing the vision had included a year to address areas of concern.

Shaw said the following year the facilitators trained approximately 400 middle and high school teachers. Initial development and training conducted over a two-day period focused on two central ideas—constructing explanations from evidence and student performance, based on the framework and NGSS. Facilitators at the Science Academies structured sessions to engage teachers as they would if they were students. Teachers followed the Gather, Reason, Communicate sequence, and engaged in writing, speaking, and producing other visual representations in order to make their thinking visible to others. Teachers participating in the training also had opportunities to work in small and large groups; engage in sustained, silent writing; develop models; and report out to their colleagues. Approaches to facilitating discourse detailed in *Ready, Set, SCIENCE!* (National Research Council, 2008) proved to be useful for the facilitators; however, Shaw indicated that they learned through their initial experiences with the Science Academies that substantial time must be devoted to preparing facilitators. In particular, the facilitators themselves have been prepared in their careers to favor content over practices, and therefore, attention must be devoted to fostering the ability to listen and to model appropriate practices throughout the training. Future trainings will offer additional sessions for middle and high school science teachers, and also expand to elementary teachers. However, Shaw cautioned that the elementary training will directly tie to CCSS for ELA and the framework for K-12 science education, rather than also addressing NGSS, because the state is first building a vision around the research in science education and not new standards. A key goal for elementary teachers is to help to get them engaged with science, since so little time is currently devoted to the subject at that level.

The second part of the South Dakota efforts to improve science education is the development of supplemental trainings for the literacy in science components of CCSS for ELA. Shaw has found that many teachers in the state are confused about what the standards mean and who is responsible for addressing them. To develop these trainings, they first examined Practice 8 of the Science and Engineering Practices in conjunction with the CCSS for ELA literacy in science standards in terms of inputs and outputs. As these trainings began, a small subset of 10 high school science teachers was invited to meet, to bring a text that they would typically use in their classrooms, and to bring a lesson plan. With guidance, teachers reviewed and compared their lesson plans to identify instances where students were gathering information, reasoning, or communicating information. Through this process, they determined that across the lesson plans, students were gathering information 80 percent of the time. These data prompted trainers to focus on how to develop units where students had more opportunities to reason and lessons placed a greater emphasis on sense-making and constructing explanations. When they examined the texts that teachers had brought, they determined through discussion that textbooks were generally ill-suited to engaging in multiple scientific and engineering practices to construct explanations of scientific phenomena. They explored ways that teachers can build in more opportunities and offer strategies that help students engage with scientific texts that do support the practices.

The literacy in science supplemental trainings consisted of a one-day meeting that focused on the shifts in practice needed and strategies that are aligned with these shifts, Shaw explained. Teachers engaged in a professional learning community online that focuses on adjusting lesson plans, implementing these plans, and reflecting on practices. Future plans for expanding this effort potentially involve making use of the Literacy Design Collaborative modules for science.

Shaw described one example of a strategy in South Dakota to help teachers learn to focus on literacy in science. They begin with examining a picture that depicts a science idea. Next, teachers are invited to write three observations, two claims that they can support with evidence and reasoning, and one aspect about which they would like to gather more evidence. After this exercise, teachers share their ideas and discuss how an activity like this can be paired with close reading of informational texts. Teachers learn how to use a text and also how to return to it repeatedly for more information and evidence. Shaw indicated that helping teachers transition from thinking about student thinking in terms of Bloom's taxonomy (Bloom, 1956) to Webb's depth of knowledge model (Webb, 2002) continues to be a challenge in their state.

Finally, Shaw shared his plans for continuing to advance science education in South Dakota. He indicated that he will continue to develop a workgroup to look for options for new standards and to make a recommendation to the South Dakota State Board of Education, while continuing to implement the framework in such a way to facilitate later transitioning to the new South Dakota science standards over the next two years. Teachers will continue to receive support related to the CCSS in reading and listening, as well as writing and speaking in science.

WORKING ACROSS STATES TO SUPPORT ENGLISH LANGUAGE LEARNERS IN LITERACY FOR SCIENCE

Okhee Lee, New York University, described work that she and her colleagues have conducted to understand what challenges and opportunities the NGSS will present to students who are ELLs. As an initial step, she, Helen Quinn, and Guadalupe Valdez engaged in an analysis of the nature of language demands in a science classroom. Lee approached the task as a science educator, Quinn as a scientist, and Valdez as an expert on language acquisition. Together, they developed a framework that described the analytical science tasks, the receptive language functions, and the productive language functions that comprised each of the NGSS practices. Next, they examined the features of the classroom language and what each of those features required teachers and students to do both orally and in writing, receptively and productively. They included the features of modality, whether communication was in small groups, to the whole class, or one on one, and registers, which described whether the task required colloquial, classroom, or disciplinary language. The framework included specific examples of registers and tasks.

Lee indicated that the analysis of the language demands of the science classroom generated interest in developing a broader framework that states could use to learn how to meet the needs of ELL students as they implement the CCSS for ELA and the NGSS. Thus, the English Language Proficiency Development (ELPD) Framework (Council of Chief State School Officers, 2012) was designed to communicate to ELL stakeholders the practices that ELLs needed to acquire for academic learning, as well as provide guidance about how to create and evaluate ELPD standards using the expectations of CCSS for ELA and NGSS as tools.

Lee then described a third initiative that had emerged from the development of the ELPD framework. She and her colleagues are working with the Council of Chief State School Officers to develop a set of standards for English language proficiency. She stated that 21 states so far would like to use the framework to develop policies to meet the needs of ELLs and promote these students' success in school. Such standards would emphasize and elaborate on the language, language knowledge, and skills using language of the CCSS for ELA, CCSS for mathematics, and the NGSS. The K-12 Practices Matrix is one element of this process that helps identify the practices in each subject and how they correspond to English language proficiency standards (Council of Chief State School Officers, 2012). A fourth initiative has recently begun to develop a teacher's guide for mathematics and science resources and how to use the resources within the ELPD framework.

Lee summarized what these three initiatives represent in her view. First, conceptually, the analyses help to tease out the language practices and functions within each of the subject areas and their corresponding standards. Second, language within the disciplines is conceptualized in terms of receptive and productive functions. Finally, this way of thinking about language is helping states develop policies and practices that can benefit all students, and especially ELLs.

DISCUSSION

Changing visions and practices in K-12 science education on larger system-level scales presents unique challenges, as several presenters pointed out. One particular challenge, working in cross-disciplinary teams, was the subject of panel discussion. Purohit of the New Visions network described differences in how the different disciplines use evidence and how the disciplines expect students to support claims with evidence in writing. In English language arts (ELA) and social studies, she noted, showing evidence in writing often means increasing length and adding more details. However, in science, evidence is more closely tied with deep reasoning and connecting observations with reading. This yields differences in what the various disciplines see as constituting a significant piece of writing. These differences emerged through conversations in cross-disciplinary teacher groups, and indicated that conversations needed to address these differences. Santos indicated that OUSD facilitates cross-disciplinary conversations at high levels of the administration, holding monthly meetings of supervisors of principals and specialists in ELA, mathematics, and science. They hold facilitated meetings focusing on the nature of academic discussion, evidence, and talk. Santos stated that they specifically take time to work in a continuous improvement process with their partners and reflect on their instructional strategies. Lastly, Purohit added that teachers across disciplines can often find common ground by focusing on the students themselves, particularly those who are struggling, more than on any particular practice.

Another point of discussion focused on finding fiscal support for these systemic initiatives. Santos indicated that, in part, science initiatives in OUSD are funded because they have become budget priorities; however, she also noted that grant writing, university partnerships, and foundation support have also been helpful. She added that these various elements must be coordinated and coherent in order to be effective. Santos also suggested that many mechanisms for sharing information and tools exist in Web-based formats, such as massive open online courses, for areas that have fewer partnership opportunities than Oakland.

One participant identified a need for greater dissemination of information and awareness of new practices in many areas of the country, perhaps occurring in regional meetings. Sam Shaw noted that he had benefited from his engagement with the Council of Chief State School Officers in this regard. However, he suggested that identifying gaps in knowledge and resources, so that support could be need-based, might prove more beneficial than targeting support by geographic area. Finally, Lee pointed out that these examples from the network, district, and state levels indicate that change can and is happening.

7

FINAL REMARKS

Many participants and presenters during the workshop noted that the Common Core State Standards for English Language Arts (CCSS for ELA) and the Next Generation Science Standards (NGSS) point to the importance of equipping students with a set of tools for critical thinking and making sense of the world, and building new knowledge upon prior knowledge in a cohesive manner. In language arts, teachers develop the means by which students can access, understand, produce, and communicate language for many purposes. In the study of science, students and teachers engage in science and engineering practices and apply crosscutting concepts to deepen their understanding of each field's disciplinary core ideas. The aspects of literacy in science included in the CCSS for ELA and the NGSS intersect and point to a new direction for instruction. The intersection also presents both challenges that need to be addressed and opportunities for mutual benefit at the classroom and school level, through professional development, and on a larger scale across systems. Many participants stressed over the course of the workshop that reading, writing, and language are natural and authentic parts of doing science. Similarly, literacy is a powerful tool for reasoning and making sense of the world. Thus, they said, reading, writing, and language do not merely overlap with science but are woven throughout all of the disciplines in school.

This chapter includes an overview of the themes and messages that emerged during the workshop, as well as steps suggested by individual participants to catalyze action.

TEXT AND TALK

Several participants presented their ideas about the nature of text and talk in science and how teachers can help students engage in the sense-making process through these means. Catherine O'Connor, Jonathan Osborne, Ann Palincsar, and Mary Schleppegrell addressed issues around text, and Okhee Lee and Sarah Michaels offered presentations focused on scientific discourse in the classroom. According to these presentations, reading, writing, and well-structured talk are all natural and critical parts of engaging in science. The unique ways that scientists write and talk requires that teachers not only give their students tools and scaffolds to help them master challenging language but also the necessary time to grapple with these challenges. Helping students develop effective reading, writing, and discourse skills means that teachers must possess key knowledge, pedagogy, and strategies, they noted.

Several presentations focused on science texts. These presenters pointed out that text plays an important and authentic role in the classroom at all levels of K-12 education. The texts used in K-12 classrooms include a variety of sources and forms, including the textbook, but also journal articles, popular magazines addressing science topics, reference materials, and Web content. As these presenters pointed out, each form of science texts is unique and often challenging for students to comprehend. The texts often include unique sentence constructions and contain words that differ from everyday and even general academic language. Science texts can be more complex and dense than other types of texts that students encounter, and are generally multimodal, including prose, but also graphs, tables, and other pictorial representations. Several presenters pointed out that because of these unique features of science texts, teachers need to support students in making sense of what they read. Specific scaffolds do exist for this purpose. However, as O'Connor stressed, students also need time to grapple with these challenging texts and to see their teachers engaging in these productive struggles.

The discourse that happens in the classroom is also important to the sense-making process. Michaels pointed to the importance of making thinking public as a means for learning how to construct scientific explanations. Just as scientific text is unique from other texts, so too is the talk. Lee suggested that scientific talk is "precise, explicit, and complex." Gaining access to this language and way of speaking is essential for all students, and is facilitated by a close look at the receptive and productive practices and ways of speaking required. Lee offered such a framework at the workshop. However, as Michaels argued, these unique ways of speaking in science about ideas will require a major shift in strategies away from the traditional Initiation-Response-Evaluation model (Mehan, 1979) that most teachers experienced themselves and learned through their preparation. To support teachers in this shift, teachers need "talk moves," she said. In her view, this high-leverage strategy is essential to successful implementation of both the CCSS for ELA and the NGSS.

CURRICULUM EXEMPLARS

The workshop included four specific examples of curricula that are successfully addressing literacy for science in different age-level settings. In addition to describing the development and key elements of these approaches, teachers implementing each of these curricula presented specific illustrations from their classrooms and their views on their experiences. Across the presentations, presenters identified a number of themes. First, these curricula demonstrate that curricula that successfully integrate science and literacy currently exist and are being effectively implemented across various levels of K-12. Importantly, a number of presentations included data that show that student outcomes in both science and literacy can improve as a result of using these curricula, including outcomes among diverse student populations. Longitudinal data presented showed that in some cases, benefits of the curriculum can persist for years beyond use of the curricula.

Jacqueline Barber described *Seeds of Science/Roots of Reading*, a science curriculum for elementary students. The model of this curriculum is based on four key components—Do, Talk, Read, and Write. The curriculum pairs first-hand and second-hand investigations centered around answering a question about a scientific phenomenon. The need to answer a scientific question gives each activity purpose, and the overarching goal is to help students be able to construct scientific explanations and make arguments. Teachers support this process through scaffolding that they fade over time. Sherrie Roland shared detailed illustrations of her implementation, emphasizing that integrating literacy for science is not only possible for typical students who are on or above grade level, but is also possible for students who are English language learners and students with special needs.

Nancy Romance described *Science IDEAS*, a curriculum developed for older elementary students. This curriculum is used in place of the language arts block and provides the content about which students read and write. The curriculum places scientific concepts at the center from which the practices emanate. Propositional concept maps are a key element of *Science IDEAS*, and they are used to build on what students already know and show how new concepts build on this knowledge and relate to each other. Students engage in hands-on investigation paired with reading, science journaling, and other activities to build and revise their maps about key science ideas over the course of a unit with teacher support. Students engage in problem solving and reflection. She said a key purpose of this approach is to help students build on their understanding of the world in a cohesive manner. Evaluation of *Science IDEAS* has indicated that this approach benefits students in both language arts and science and that these benefits are long-lasting.

LeeAnn Sutherland described *Investigating and Questioning our World through Science and Technology (IQWST)*, a science curriculum for middle school students. Designed to help foster cohesion both within and across physical science, chemistry, earth science, and life science, *IQWST* seeks to build students' skills in sense-making and using claims, evidence, and reasoning to develop explanations for driving questions. The curriculum is organized into 8- to 10-week units, during which students engage in investigations, reading, writing, and talking. Deborah Peek-Brown shared her observations about *IQWST* in the classroom. In her view, learning to use the language of argumentation is more than a way of talking or writing; it is a way of thinking about the world. *IQWST* uses a supported, iterative process to help students learn to communicate their thinking to others.

Susan Goldman and Cynthia Greenleaf shared their perspectives on *Project READi*, a curriculum for students at the middle and high school levels. Greenleaf shared that this curriculum is designed to supplement an existing curriculum that includes hands-on investigation. It consists of text-based modules composed of carefully selected and sequenced readings that help students develop causal models of explanations of scientific phenomena. Students are encouraged to grapple with the evidence presented within and across texts to be able to answer questions, like "How do we know?" or "How do we make sense of our differences?" A key aspect of *Project READi* is helping students to shift their way of approaching reading in science to one of active engagement and orientation to looking for evidence and support for claims. Greenleaf shared that this approach has been effective but has required adequate support for teachers.

PREPARING TEACHERS

The workshop included five case studies that focused on how to prepare teachers to teach science in ways that help students to make sense of the world, addressing the particular element of how to prepare them to integrate literacy for science. Two of these cases described ways to prepare preservice teachers, and three cases focused on professional development for practicing teachers.

Across these presentations, several common themes emerged. First, these approaches for preparing teachers emphasized engaging teachers in experiences that mirror those that their students should be engaging in, including working in communities, engaging in hands-on activities, reading, and writing. Many presenters emphasized helping teachers become aware of their own strategies for making sense of text. Constructing a culture for learning where it was acceptable to ask questions and make mistakes proved important for teachers as well as for their students, according to these presenters.

Each of the cases had unique aspects as well. Elizabeth Davis emphasized that novice teachers need not only to gain knowledge of disciplinary content but also knowledge of pedagogy, ways that students learn and common difficulties they encounter. Teachers need to be equipped with a set of strategies and many opportunities to practice them. Mark Windschitl emphasized his program at the University of Washington focuses on four core strategies—planning for engagement, eliciting student ideas, supporting ongoing changes in thinking, and pressing for evidence-based explanations. He indicated that their students learn how to plan units around compelling, anchoring events as they learn to enact science practices. Novice teachers engage first in short practice opportunities and move toward longer and more complex practice experiences with students. Attention is given to how to elicit student ideas and help them work toward building evidence-based explanations for scientific phenomena.

Jean Moon described Next Generation Science Exemplar-Based Professional Learning Systems, which makes extensive use of video exemplars to help teachers see what is possible and what literacy for science can look like. Teachers across K-12 meet face to face and engage with Web-based content as they build their knowledge and strategies.

Two other case studies illustrated a more sustained level of professional development. In the case of Quality Teaching for English Learners at the International Newcomers' Academy, Aida Walqui described how her organization worked with the school over a period of three years to help build teacher skill and knowledge, as well as teacher capacity in the form of training in-house coaches and facilitators, and understanding and support among administrators. Key to their support of students just beginning to learn English is a "pedagogy of promise," where students are viewed as apprentices as they learn the language of school and science through engagement in the science practices. Brett Moulding described Partnership for Effective Science Teaching and Learning, another three-year professional development program with a primary focus on science performance. In this approach, teachers engage in a process that mirrors that of their students and learn to think about science concepts in relation to causality, patterns, and systems, and to construct explanations through gathering, reasoning, and communicating.

SUPPORT FOR LITERACY FOR SCIENCE ON VARIOUS SCALES

The workshop featured several examples of support for literacy for science occurring across multiple schools in a network, district, or state. The purpose of these presentations was to learn from the experiences of those working to make systemic change to K-12 science education. Although the approaches varied greatly based on location, scope, and goals, these larger scale supports yielded several common aspects. First, broader efforts to bring about changes in teacher knowledge, approaches, and strategies involved fostering communities of learning and strategies that emulated approaches to be used with students. However, effecting these changes began with working to create a shared vision for science education, and more specifically literacy for science. This process required time and engagement with individuals at all levels of the system, and was often facilitated by supportive policies at the district or state levels. Often changes were phased in. Several presenters described their efforts to build capacity in their systems to support and sustain the changes through training trainers, working with principals, or supporting "trailblazing schools," among other strategies. Finally, a multistate effort is under way to consider the needs of English language learners as the CCSS and NGSS are implemented. Each of the individual approaches to supporting science education on a larger scale is summarized below.

Kiran Purohit of New Visions for Public Schools described her experiences working with one New York charter school, New Visions High School for Advanced Math and Science. New Visions supports a network of public and charter schools across New York City. A key aspect of their support involved implementing the Literacy Design Collaborative, which supports reading and writing across disciplines. One key feature that Purohit shared with the participants at the workshop included innovative ways to engage community experts throughout science investigations beginning with the planning phase. Cross-disciplinary teacher groups are also central to their approach and help to focus teachers around evaluating writing and their overall needs as students.

Maria Santos shared her experiences in leading systems change at the Oakland Unified School District (OUSD), a large, urban district with many low-income students. Santos described the system-wide plan that they have begun implementing in phases. Supported by key policies and funding, OUSD identified three goals related to literacy for science that they have worked to address: close reading of complex text, academic discussion, and evidence-based writing. Santos stressed the importance of engaging all adults at all levels of the system to build a shared vision. They have also used many vehicles and tools to enact change, including fostering "trailblazer schools," establishing summer institutes, and providing extensive professional development to both teachers and administrators. OUSD also engages principal supervisors and specialists in dialog and facilitated sessions about science education as well.

Sam Shaw discussed his experiences leading state-wide change in science education in South Dakota. He described three initiatives, bolstered by their governor's package to support teachers, to create a shared vision for science education based on the framework. The first of these initiatives was the creation of Science Academies where a set of facilitators and later other science teachers engaged in professional development to build understanding of new ways of teaching science. Second, they engaged teachers in supplemental training around literacy in science standards of the Common Core, ultimately working to analyze and adapt science lessons to address obtaining, as well as evaluating and communicating information. Finally, Shaw will be working to determine how his state should move forward with adoption of the new South Dakota science standards in a seamless fashion. Like other presentations, the lesson from South Dakota shows that the change process and building a shared vision across stakeholders takes time.

THEMES FROM THE WORKSHOP

David Pearson, chair of the workshop planning committee, summarized some themes that he identified on each day of the workshop. He noted that the presentations on the first day focused primarily on the conceptual issues regarding how literacy in science is portrayed in CCSS for ELA and NGSS, principles and practices important to both literacy and science, the nature of science texts and discourse, and features of science curricula that encourage the literacy in science practices called for in the standards documents. Pearson identified six themes that he said emerged to him across the range of topics discussed, and other committee and audience members expanded upon these and other ideas.

Pearson suggested that one theme that he saw emerge on Day 1 was the centrality of questions. Questions provide the reason for engaging in science, whether it is through hands-on investigation or engagement with science texts. Related to this notion is student engagement, a second, often implicit, theme noted by Pearson. He said when students have a reason to read, a reason to learn terminology, and compelling and interesting content, their level of engagement increases, as well as their stamina and cognitive effort. Student engagement is key to any learning, Pearson stated, but it has been the hallmark of project-based learning in science in particular for some time.

Coherence is a third theme he suggested. As students learn to make sense of the world, constructing causal explanations of scientific phenomena, they are building on their prior knowledge. Pearson noted that coherence should exist within science texts as well as in science curricula. Brian Reiser added that this curricular coherence is best when it is not only within a single strand (i.e., life science) but also across science strands (e.g., life science and chemistry).

Within these cultures of hands-on, problem-based learning, Pearson noted that content, literacy, and scientific practices do not merely overlap, but are integrated and interwoven throughout the disciplines. Reading, writing, and oral language are interwoven not just through science, but also mathematics, social studies, and literature. Several curriculum developers and teachers described a number of examples of the ways in which science and literacy can be effectively integrated.

Representation was, to Pearson, a theme that emerged from a number of presentations, a term that he stated has more than one meaning. In one sense, representation can mean mapping an icon to an idea, but it can also mean transforming information and showing an idea in a new way, such as from a verbal idea to an image or even transforming an idea from one verbal form to another verbal form. In this way, representation is at the intersection of science ideas and their integral relationship with language. Helen Quinn argued that representation, along with fostering rich language development, and reasoning and analysis form the underpinnings of any successful approach to learning. "Representation is key to learning," Quinn stated. "For students to learn to represent their own ideas through building their own models is a very important part of learning."

Teachers also need to experience the struggles that their students experience in reading and writing, Pearson commented. Although some teachers can relate to their students who struggle with writing, few go through experiences that enable them to empathize with their students who struggle with reading, according to Pearson. He and his colleagues have found that when teachers are forced to engage with a highly challenging text in a training situation, they become more attuned to the need to build an infrastructure of strategies for making sense of difficult texts for their students.

In Pearson's view, another key conceptual issue that emerged through discussion was the transfer of authority in the classroom. Sarah Michaels and Susan Pimentel both stressed the need for teachers to open discussion in the classroom, he noted, and to "let go" more, letting students lead the discussion to a much greater degree. As Michaels stated, "Kids are much more powerful learners than we've let them be." The purpose is to help students look to the evidence and to use their discourse to adjudicate disagreements as they work to construct explanations, rather than looking to the teacher for the correct answer. Many presenters noted that this requires addressing the culture of the classroom. A number of others noted that teachers need to see this occurring through video or other means to believe it is possible and to know how to elicit this type of discourse.

Helen Quinn shared her perspective that hands-on investigations and engagement with text is not an "either/or" proposition, and that students need to be using many types of texts. However, she suggested that these shifts in strategies require a lot of teachers, and that translating research to effective professional development for teachers is a big challenge for the field. Juan-Carlos Aguilar said he shared that concern and argued that care be taken to communicate clearly to teachers what the standards do and do not require of them.

The second day of the workshop addressed the supports that are needed to help teachers implement literacy for science practices in the classroom, including the necessary professional development, as well as the administrative and systemic supports. Pearson and several other participants summarized some issues that they said had emerged, and they and other individual audience members suggested a few areas that could be addressed through policy and research.

One of the main supports that teachers need is time, Pearson stated. Repeatedly during the workshop, participants stated that students and teachers need time to engage in productive struggle, grapple with challenging science texts, and explore content deeply. They need time for sense-making and the revision process. Pimentel noted that a big message for her was the need to slow down in the classroom. Elizabeth Moje shared her view that the field needs to help teachers learn how to engage students in more "doing" of science within the existing structures and time constraints of schools, particularly secondary schools where students move from class to class. In addition to this classroom time, teachers need time to develop new strategies and they, along with administrators, need time to adjust to this new vision. Some participants indicated that this means giving teachers space away from high-stakes testing.

Pearson noted that the need for scaffolding emerged across many presentations. As he stated, scaffolding "is a part of our DNA as a profession." These supports are needed for both teachers and their students, collectively and individually. These scaffolds are enabled by structure, but they are flexible and responsive. For both teachers and students, guidance helps with the acquisition of knowledge and practices. In regard to professional development, he said it is clear that teachers need approaches that mirror what should be happening with students in communities of practice. He also noted that scaffolds and strategies that teachers use must be flexible, rather than rigid routines. Another participant also cautioned against one-size-fits-all approaches, citing recent Program for International Student Assessment results that showed that as teachers had more autonomy in the classroom, their students' test scores improved. He argued for an approach where groups of teachers work together to become "cogenerators and coreflectors on the instruction that they are trying to do."

Pearson suggested that as teachers work to support classroom discourse and other literacy practices in service of science, they need a range of exemplars to use, particularly through video. According to Moje, videos can help elucidate complex practices. They are especially needed at the high school level and specifically related to literacy for science, in her view. She and others added that collections of videos need to be accompanied by descriptions and explanations, address a range of grade levels and developmental progressions, and show that adaptation and individualization is possible. Teachers also need access to a range of resources and tools, and to build their skills in differentiating instruction beyond access to video exemplars. Michaels shared her view that the CCSS for ELA and NGSS provide great opportunities for sharing resources and knowledge in the form of exemplars of great practice to achieve common goals.

IDEAS FOR POLICY AND RESEARCH

The final phase of the workshop asked panelists and participants to consider the policy implications of the topics discussed over the two days of the workshop, including potential barriers and supports from policy. Several panelists and participants suggested elevating in importance the profile of science in K-12 education to be more on par with literacy and mathematics at the school, district, state, and national levels to enact change on a wider scale. Quinn suggested that the workshop had offered "existence proofs," showing that professional development can effectively shift teacher strategies in ways that demonstrably improve student learning. She and others suggested that policies that can create "space" for teachers to learn new instructional strategies can be a necessary and important opportunity.

Jacqueline Barber indicated that, in her view, the NGSS has created an opening to elevate literacy for science at the policy level. Moreover, she suggested that existing data and policies can be used to engage policymakers to create supports for literacy for science in schools. "Don't reinvent the policy wheel," she cautioned. Others said they would like to see additional exploration of how to systematically change teacher strategies on a large scale without watering down the vision of the NGSS, along with a timeline for changes on this scale.

A number of practical and conceptual topics related to literacy for science in K-12 education were beyond the scope of what could be addressed in a two-day workshop. A primary topic that participants raised more than once was assessment. Pearson related his concern that attaching high stakes to assessment is problematic regardless of the method of assessment used. He suggested that the field needs to continue to push for better links between assessments, standards, and practices. Others had remaining questions about whether there should be separate assessments for literacy and science with "leakage" or a common assessment at the overlap, as well as how teachers would be evaluated on literacy for science practices.

The nature of the relationship between literacy and science was a topic that many suggested needs additional study. How the nature of evidence, argumentation, and explanation may differ across the disciplines of language arts and science is still an open question, in Pearson's view. However, Pimentel added that the emphasis of claims, reasoning, and evidence in the CCSS for ELA is consistent with the similar emphases in the NGSS. In addition, she said, future research could address in more depth if and how practices and benefits in one discipline transfer to the other. For example, she asked, "If I focus on meaning-making during science, will my students get better at meaning-making in reading comprehension?" Similarly, research and further discussion can continue the conversation about the degree to which various tools, strategies, and practices can or should travel across disciplines. Others felt that more research was needed on the impact of literacy for science on ELA outcomes. More conversation is needed to clarify the roles of both teachers of science and ELA teachers, Pearson suggested. Moje offered that the focus for science teachers should be on science literacy only, as distinct from ELA more generally.

Finally, many panelists and presenters commented that innovations and research need to reach the field through professional development and other communications. Pimentel stated, "Despite the challenges that exist to integrating science investigations and literacy for science on a large scale, I leave here really invigorated . . . because it shows what is possible. It shows that we need not keep high-level thinking . . . about content from any of our students." As she summarized, presentations at the workshop showed that teachers must believe in the abilities of their students, and that all students are capable of learning at a high level.

REFERENCES

ACT. (2006). *Reading between the Lines: What the ACT Reveals About College Readiness in Reading*. Iowa City, IA: Author

Anderson, J.R. (1987). Skill acquisition: Compilation of weak-method problem solutions. *Psychological Review, 94*, 192-210.

August, D., and Shanahan, T. (2006). *Developing Literacy in Second-Language Learners: Report of the National Literacy Panel on Language-Minority Children and Youth*. Mahwah, NJ: Lawrence Erlbaum Associates.

Berk, L., and Windschitl, M. (2013). *Building a Repertoire of Literacy Support and Practices*. Presentation at the Exploring the Overlap between "Literacy in Science" and the Practice of Obtaining, Evaluating, and Communicating Information: A Workshop, December 9-10, National Research Council, Washington, DC. Available: http://sites.nationalacademies.org/DBASSE/BOSE/DBASSE_086848 [July 2014].

Bloom, B.S. (1956). *Taxonomy of Educational Objectives, Handbook I: The Cognitive Domain*. New York: David McKay.

Brown, J.S. (1989). Towards a new epistemology for learning. In C. Frasson and J. Gauthiar (Eds.), *Intelligent Tutoring Systems at the Crossroads of Artificial Intelligence and Education*. Norwood, NJ: Ablex.

Cazden, C., and Mehan, H. (1989). Principles from sociology and anthropology: Context, code, classroom, and culture. In M. Reynolds (Ed.), *Knowledge Base for the Beginning Teacher*. Oxford: Pergamon Press.

Cheuk, T. (2013). *Relationships and Convergences among The Mathematics, Science, and ELA Practices. Refined Version of Diagram Created by the Understanding Language Initiative for ELP Standards*. Stanford, CA: Stanford University.

Council of Chief State School Officers. (2012). *Framework for English Language Proficiency Development Standards Corresponding to the Common Core State Standards and the Next Generation Science Standards*. Washington, DC: Author.

Davis, E.A., and Bricker, L.A. (2013). *Science and Literacy in Teacher Education: What Do Preservice Teachers Need to Know and Be Able to Do?* Paper prepared for the Exploring the Overlap between "Literacy in Science" and the Practice of Obtaining, Evaluating, and Communicating Information: A Workshop, December 9-10, National Research Council, Washington, DC. Available: http://sites.nationalacademies.org/DBASSE/BOSE/DBASSE_086022 [July 2014].

Frayer, D., Frederick, W.C., and Klausmeier, H.J. (1969). *A Schema for Testing the Level of Cognitive Mastery*. Madison: Wisconsin Center for Education Research.

Garcia, E., and Cuellar, D. (2006). Who are these linguistically and culturally diverse students? *Teachers College Record, 108*, 2220-2246.

Goldman, S.R., and Rakestraw, J.A. (2000). Structural aspects of constructing meaning from text. In M.L. Kamil, P.B. Mosenthal, P.D. Pearson, and R. Barr (Eds.), *Handbook of Reading Research* (Vol. II, pp. 311-335). Mahwah, NJ: Lawrence Erlbaum Associates.

Graesser, A.C., McNamara, D.S., and Louwerse, M.M. (2003). What do readers need to learn in order to process coherence relations in narrative and expository text. In A.P. Sweet and C.E. Snow (Eds.), *Rethinking Reading Comprehension* (pp. 82-98). New York: Guilford.

Greenleaf, C., Brown, W., Goldman, S., and Ko, M-L. (2013). *Designing at the Crossroads of NGSS and CCSS: Text-Based Investigations for Evidence-Based Argumentation in Science*. Presentation at the Exploring the Overlap between "Literacy in Science" and the Practice of Obtaining, Evaluating, and Communicating Information: A Workshop, December 9-10, National Research Council, Washington, DC. Available: http://sites.nationalacademies.org/DBASSE/BOSE/DBASSE_086307 [July 2014].

Halliday, M.A.K. (1994). *An Introduction to Functional Grammar* (2nd ed.). London: E. Arnold.

Krajcik, J.S., Reiser, B.J., Sutherland, L.M. and Fortus, D. (2011). *IQWST: Investigating and Questioning our World through Science and Technology. Middle School Science Curriculum Materials*. Greenwich, CT: Sangari Active Science.

Lee, O., and Llosa, L. (2011-2015). *Promoting Science among English Language Learners (P-SELL) Scale-up*. Washington, DC: National Science Foundation, Discovery Research K-12.

Mehan, H. (1979). *Learning Lessons: Social Organization in the Classroom*. Cambridge, MA: Harvard University Press.

Miller, D., and Purohit, K. (2013). *Reframing Science Literacy through Community Engagement*. Presentation at the Exploring the Overlap between "Literacy in Science" and the Practice of Obtaining, Evaluating, and Communicating Information: A Workshop, December 9-10, National Research Council, Washington, DC. Available: http://sites.nationalacademies.org/DBASSE/BOSE/DBASSE_086310 [July 2014].

Moulding, B. (2013). *Partnership for Effective Science Teaching and Learning: Gathering, Reasoning, and Communicating*. Presentation at the Exploring the Overlap between "Literacy in Science" and the Practice of Obtaining, Evaluating, and Communicating Information: A Workshop, December 9-10, National Research Council, Washington, DC. Available: http://sites.nationalacademies.org/DBASSE/BOSE/DBASSE_086311 [July 2014].

National Governors Association and Council of Chief State School Officers. (2010a). *Common Core State Standards for English Language Arts & Literacy in History/Social Studies, Science, and Technical Subjects*. Available: http://www.corestandards.org/wp-content/uploads/ELA_Standards.pdf [March 2014].

National Governors Association and Council of Chief State School Officers. (2010b). *Common Core State Standards for Mathematics*. Available http://www.corestandards.org/wp-content/uploads/Math_Standards.pdf [March 2014].

National Research Council. (1996). *National Science Education Standards*. National Committee on Science Education Standards and Assessment. Washington, DC: National Academy Press.

National Research Council. (1999). *How People Learn: Brain, Mind, Experience, and School*. Committee on Developments in the Science of Learning. J.D. Bransford, A.L. Brown, and R.R. Cocking (Eds.). Washington, DC: National Academy Press.

National Research Council. (2008). *Ready, Set, SCIENCE! Putting Research to Work in K-8 Science Classrooms*. S. Michaels, A.W. Shouse, and H.A. Schweingruber, Editors. Board on Science Education, Center for Education, Division of Behavioral and Social Sciences and Education. Washington, DC: The National Academies Press.

National Research Council. (2012). *A Framework for K-12 Science Education: Practices, Crosscutting Concepts, and Core Ideas*. Committee on a Conceptual Framework for New K-12 Science Education Standards. Board on Science Education, Division of Behavioral and Social Sciences and Education. Washington, DC: The National Academies Press.

NGSS Lead States. (2013). *Next Generation Science Standards: For States, By States*. Washington, DC: The National Academies Press.

Osborne, J. (2013). *What Do Science Teachers Need to Know about Language in Science?* Presentation at the Exploring the Overlap between "Literacy in Science" and the Practice of Obtaining, Evaluating, and Communicating Information: A Workshop, December 9-10, National Research Council, Washington, DC. Available: http://sites.nationalacademies.org/DBASSE/BOSE/DBASSE_086301 [July 2014].

Palincsar, A. (2013). The Next Generation Science Standards and the Common Core State Standards: Proposing a happy marriage. *Science and Children, 51*(1), 10-15.

Palincsar, A.S., Symons, C., and Schleppegrell, M. (2013). *Teaching Functional Grammar Analysis as a Means to Support English Language Learners: Interactions with Informational Text.* Paper presented at the annual meeting of the American Educational Research Association, San Francisco, CA.

Pauk, W., and Owens, R.J.Q. (2008). *How to Study in College* (9th ed.). Boston, MA: Houghton Mifflin.

Pimentel, S. (2013). *Next Generation Science Standards and the Common Core.* Presentation at the Exploring the Overlap between "Literacy in Science" and the Practice of Obtaining, Evaluating, and Communicating Information: A Workshop, December 9-10, National Research Council, Washington, DC. Available: http://sites.nationalacademies.org/DBASSE/BOSE/DBASSE_086850 [July 2014].

Putnam, R.T., and Borko, H. (2000). What do new views of knowledge and thinking have to say about research on teacher learning? *Educational Researcher, 29*(1), 4-15.

Romance, N.R., and Vitale, M.R. (2011). *Interdisciplinary Perspectives for Linking Science and Literacy: Implications from Multi-Year Studies Across Grades K-5.* Paper presented at the Annual Meeting of the American Educational Research Association, New Orleans, LA.

Sadler, P.M., Sonnert, G., Coyle, H.P., Cook-Smith, N., and Miller, J.L. (2013). The influence of teachers' knowledge on student learning in middle school physical science classrooms. *American Educational Research Journal, 50*(5), 1020-1049.

Schleppegrell, M.J. (2001). Linguistic features of the language of schooling. *Linguistics and Education, 12*(4), 431-459.

Schleppegrell, M.J. (2004). *The Language of Schooling: A Functional Linguistics Perspective.* Mahwah, NJ: Lawrence Erlbaum Associates.

Schleppegrell, M.J., and Palincsar, A.S. (2013). *What Do Teachers Need to Know about the Language of Science Text?* Presentation at the Exploring the Overlap between "Literacy in Science" and the Practice of Obtaining, Evaluating, and Communicating Information: A Workshop, December 9-10, National Research Council, Washington, DC. Available: http://sites.nationalacademies.org/DBASSE/BOSE/DBASSE_086553 [July 2014].

Stage, E.K., Asturias, H., Cheuk, T., Daro, P. A., and Hampton. S.B. (2013). Opportunities and challenges in next generation standards. *Science, 340*(6130), 276-277.

Smith, F. (1978). *Reading.* New York: Teachers College Press.

Sweet, A.P., and Snow, C.E. (Eds.) (2003). *Rethinking Reading Comprehension. Solving Problems in the Teaching of Literacy.* New York: Guilford Press.

Vitale, M.R., and Romance, N.R. (2013). *Science IDEAS: Integrated Reading/Language Arts within Science Instruction Across Grades K-5*. Presentation at the Exploring the Overlap between "Literacy in Science" and the Practice of Obtaining, Evaluating, and Communicating Information: A Workshop, December 9-10, National Research Council, Washington, DC. Available: http://sites.nationalacademies.org/DBASSE/BOSE/DBASSE_086554 [July 2014].

Vygotsky, L. (1986) Thought and language: Revised edition. *Journal of Mind and Behavior, 8*, 175-177.

Walqui, A., Warren, T., and Ruiz, J. (2013). *The Development of Scientific Literacies with English Language Learners: The INA-QTEL Collaboration*. Presentation at the Exploring the Overlap between "Literacy in Science" and the Practice of Obtaining, Evaluating, and Communicating Information: A Workshop, December 9-10, National Research Council, Washington, DC. Available: http://sites.nationalacademies.org/DBASSE/BOSE/DBASSE_087376 [July 2014].

Webb, N. (2002). *Depth-of-Knowledge Levels of Four Content Areas*. Unpublished paper. Available: http://facstaff.wcer.wisc.edu/normw/All%20content%20areas%20%20DOK%20levels%2032802.pdf [June 2014].

A

WORKSHOP AGENDA

EXPLORING THE OVERLAP BETWEEN "LITERACY IN SCIENCE" AND THE PRACTICE OF OBTAINING, EVALUATING, AND COMMUNICATING INFORMATION: A WORKSHOP

National Academy of Sciences Building
Lecture Room
2101 Constitution Ave, NW
Washington, DC 20418
December 9-10, 2013

Webcast
http://sites.nationalacademies.org/DBASSE/BOSE/CurrentProjects/DBASSE_083999

December 9, 2013

8:30–8:45 AM **Registration, Coffee, Networking**

8:45–9:00 AM **Welcome**

- ***Heidi Schweingruber,*** Deputy Director, Board on Science Education
- ***David L. Evans,*** Executive Director, National Science Teachers Association

9:00–9:30 AM **Connecting Literacy for Science in New ELA and Science Standards**
- ***David Pearson,*** Study Chair, University of California Berkeley
- ***Helen Quinn,*** Committee Member, Stanford University

SESSION 1: OVERVIEW OF LITERACY FOR SCIENCE IN THE STANDARDS

9:30–10:00 AM **K-12 Education Standards and Literacy in Science**

Moderator
- ***Juan-Carlos Aguilar,*** Committee Member, Georgia Department of Education

Speakers
- ***Brian Reiser,*** Northwestern University
- ***Sue Pimentel,*** Committee Member, Student Achievement Partners

10:00–10:30 AM **Audience Think-Pair-Share**

Facilitator
- ***Catherine O'Connor***, Boston University
- ***Sarah Michaels,*** Committee Member

10:30–10:40 AM **Break**

SESSION 2: LITERACY FOR SCIENCE IN PRINCIPLE AND PRACTICE

10:40–11:25 AM **Role of Literacy for Science: Panel Discussion**

Discussants
- ***Sarah Michaels,*** Committee Member, Clark University
- ***David Pearson,*** Committee Member, University of California, Berkeley
- ***Elizabeth Birr Moje,*** Committee Member, University of Michigan

11:25–12:25 PM **What Do Teachers Need to Know about the Language of Science Text**

Moderator
- ***Helen Quinn,*** Committee Member

Speakers

- ***Jonathan Osborne,*** Stanford University
- ***Mary Schleppegrell*** and ***Annemarie Palincsar,*** University of Michigan
- ***Catherine O'Connor,*** Boston University

12:25–1:20 PM	**Poster Session: Participants from the Research-Based, Standards Aligned Instruction Session** *(Lunch Served)*
1:20–2:20 PM	**Research-Based, Standards Aligned Instruction** Moderator - ***Elizabeth Birr Moje,*** Committee Member Speakers - ***Jacqueline Barber,*** Seeds of Science/Roots of Reading *and* ***Sherrie Roland,*** Grafton Village Elementary School - ***Nancy Romance*** and ***Michael Vitale,*** *Science IDEAS* - ***LeeAnn Sutherland*** and ***Deborah Peek-Brown,*** *Investigating and Questioning our World through Science and Technology*
2:20–2:30 PM	**Break**
2:30–3:15 PM	**The Role of Oral and Written Discourse in Teaching and Learning Science** Moderator - ***Sue Pimentel,*** Committee Member Speakers - ***Susan Goldman*** and ***Monica Ko,*** University of Illinois at Chicago; ***Cynthia Greenleaf*** and ***Willard Brown,*** WestEd - ***Okhee Lee,*** New York University

SESSION 3: WRAP UP DAY 1

3:15–4:15 PM	**Bringing It Together: Audience Discussion** Facilitator - ***Catherine O'Connor,*** Boston University - ***Sarah Michaels,*** Committee Member

4:15–5:00 PM	**Major Themes and Lessons Learned** Moderator - ***David Pearson,*** Committee Chair Speakers - ***Committee Members*** - ***Audience***
5:00 PM	**Adjourn**

December 10, 2013

8:30–9:00 AM	Food and Coffee

SESSION 4: SCIENCE LITERACY DEMANDS ON THE EDUCATION SYSTEM

9:00–11:00 AM	**Teacher Preparation and Professional Development** Moderator - ***Helen Quinn,*** Committee Member Research - ***Elizabeth Davis,*** University of Michigan Case Studies - ***Mark Windschitl,*** University of Washington *and* ***Lindsay Berk,*** Chinook Middle School in Sea Tac - ***Jean Moon,*** Next Generation Science Exemplar-Based Professional Learning System and ***Jocelyn Lloyd,*** Woodland School in Worcester - ***Aida Walqui,*** WestEd *and* ***Tanya Warren,*** International Newcomer Academy - ***Brett Moulding,*** Partnership for Effective Science Teaching and Learning
11:00–11:10 AM	**Break**
11:10 AM–12:25 PM	**Models for Supporting Literacy for Science** Moderator - ***Juan-Carlos Aguilar,*** Committee Member

Case Studies

- ***Kiran Purohit*** and ***Danielle Miller***, New Vision Schools
- ***Maria Santos***, Oakland Unified School District
- ***Sam Shaw,*** South Dakota Department of Education

12:25–1:10 PM **Poster Session: Participants from the Teacher Preparation and Professional Development Session** (*Lunch Served*)

SESSION 5: WHAT THIS MEANS FOR YOU

1:10–1:55 PM **Bringing It Together: Audience Discussion**

Facilitator

- ***Catherine O'Connor***, Boston University
- ***Sarah Michaels,*** Committee Member

SESSION 6: WRAP UP DAY 2

1:55–2:50 PM **Major Messages, Lessons Learned and Next Steps**

Moderator

- **David Pearson**, Committee Chair

Speakers

- ***Committee Members***
- ***Audience***

2:50–3:00 PM **Wrap-up**

Speakers

- ***David Pearson,*** Committee Chair
- ***Michael Feder,*** Workshop Director

3:00 PM **Adjourn**

B

REGISTERED WORKSHOP PARTICIPANTS

Last Name	First Name	Affiliation
Aguilar	Juan-Carlos	Georgia Department of Education
Autrey	Jessica	Quality Teaching for English Learners
Bailin	Deborah	Union of Concerned Scientists
Baker	Tanya	National Writing Project
Barber	Jacqueline	Lawrence Hall of Science
Bartiromo	Margo	Merck Institute for Science Education
Baxter	Gail	Educational Testing Service
Bell	Rita	Monterey Bay Aquarium
Bell	Mary	Prince George's County Public Schools
Bennett	Theresa	Delaware Department of Education
Bernhardt	Anita	Maine Department of Education
Blaha	Leslie	Montgomery Blair High School, Maryland
Brown	Willard	WestEd
Byers	Albert	National Science Teachers Association
Campbell	Lynne	Iowa STEM Initiative / Iowa State University
Carlson	Lindsay	Highline Public Schools
Cheuk	Tina	Stanford University, Understanding Language
Crosby	Greg	U.S. Department of Agriculture
Curless	Melinda	Kentucky Department of Education

Last Name	First Name	Affiliation
Davis	Carmen	Red Clay Consolidated School District
Davis	Betsy	University of Michigan
Davy	Lucille	The Hunt Institute
Dieterle	Edward	Gates Foundation
Duschl	Richard	National Science Foundation
Earle	Janice	National Science Foundation
Eliopoulos	Teresa	Achieve
Emrick	Thomas	Smithsonian
Eros	Terri	Red Clay Consolidated School District
Evans	David	National Science Teachers Association
Ford	Michael	University of Pittsburgh
Gabel	David	Consultant
Galyas	Lesley	New Mexico Public Education Department
Goldman	Daniel	Rosa Parks Middle School, Maryland
Goldman	Susan	University of Illinois, Chicago
Greenleaf	Cynthia	WestEd
Griffin	Darion	American Federation of Teachers
Gruber	Stacey	Merck Institute for Science Education
Hain	Bonnie	Baltimore County Public Schools
Harrison	Molly	National Oceanic and Atmospheric Administration
Hedges	Gary	Maryland State Department of Education
Henriquez	Andres	National Science Foundation
Higgins	Melinda	Triangle Coalition for STEM Education
Hill	Jerome	Red Clay Consolidated School District
Hobbs	Melanie	American Federation of Teachers
Horak	Jennifer	National Science Teachers Association
Jenkins	Larinda	Alexis I. DuPont Middle School
Johnson	Pat	U.S. Department of Education
Joslin	Julie	North Carolina, Department of Public Instruction
Kapur	Ambika	Carnegie Corporation of New York
Kent	Saundra	New Hampshire Department of Education
Ko	Monica	University of Illinois at Chicago

Last Name	First Name	Affiliation
Koch	Louisa	National Oceanic and Atmospheric Administration
Kochhar-Bryant	Carol	George Washington University
Koppal	Mary	American Association for the Advancement of Science
Lee	Sera	Arlington Public Schools, Virginia
Lee	Okhee	New York University
Lewenstein	Bruce	Cornell University
Lloyd	Jocelyn	Worcester Public Schools, Massachusetts
Marshall	Sandy	Sandra Marshall & Associates
Massado	Karon	Red Consolidated School District
Matsumoto	George	Monterey Bay Aquarium Research Institute
Mayville	Melissa	National Education Association
McGrath	Edward	Red Clay Consolidated School District
McPhillips	Janel	Calvert County Public Schools, Maryland
Mead	Tonyea	Delaware Department of Education
Merrill	Margaret	Merrill Educational Consulting
Meyer	Melinda	New Canaan Public Schools
Michaels	Sarah	Clark University
Miller	Zipporah	Anne Arundel County Public Schools, Maryland
Miller	Danielle	New Visions for Public Schools
Milligan	Stephanie	Howard County Public School System
Moje	Elizabeth Birr	University of Michigan
Moncure	Clark	Red Clay Consolidated School District
Moon	Jean	Tidemark Institute
Moravchik	Bruce	National Oceanic and Atmospheric Administration
Morrison	Jan	Teaching Institute of Excellence in STEM
Moulding	Brett	Partnership for Effective Science Teaching & Learning
Moyer	John	Delaware Department of Education
Noun	Judith	New Teacher Center eMSS
O'Connor	Catherine	Boston University
Opperman	Julianne	Maine School Administrative District 51
Osborne	Jonathan	Stanford University
Palincsar	Annemarie	University of Michigan

Last Name	First Name	Affiliation
Pearson	David	University of California, Berkeley
Peek-Brown	Deborah	University of Michigan
Pimentel	Susan	Student Achievement Partners
Pruitt	Stephen	Achieve
Purohit	Kiran	New Visions for Public Schools
Quinn	Helen	Stanford University, emerita
Rapier	Becky	Arizona Department of Education
Reese	Amy	Howard County Public School System, Maryland
Reiser	Brian	Northwestern University
Renzulli	Andrew	Harford County Public Schools
Rieth	Mary Beth	Mount Notre Dame High School
Roberts	Joann	Maryland State Department of Education
Robinson	Matthew	Red Clay Consolidated School District
Roland	Sherrie	Stafford County Public Schools
Romance	Nancy	Florida Atlantic University
Rossiter	Dione	American Association for the Advancement of Science
Ruiz	Javier	WestEd
Ryu	Minjung	Johns Hopkins University
Santos	Maria	Oakland Unified School District
Schatz	Dennis	National Science Foundation
Schleppegrell	Mary	University of Michigan
Sergovic	Karen	Esperanza Academy
Shaw	Sam	South Dakota Department of Education
Sikorski	Tiffany-Rose	George Washington University
Simani	Maria	California Science Project
Smith	Scott	Idaho State Department of Education
Sneideman	Joshua	U.S. Department of Energy
Solomon	Gerald	Samueli Foundation
Sosa	Maria	American Association for the Advancement of Science
Steffen	Peggy	National Oceanic and Atmospheric Association
Stricklin	Andrew	Fort Worth Independent School District
Sutherland	LeeAnn	University of Michigan

Last Name	First Name	Affiliation
Thurston	Suzanne	American Association for the Advancement of Science
Tierney	Susan	Measured Progress
Tupas	Jermelina	National Science Foundation
van der Veen	Wilhelmus	Raritan Valley Community College
VanderPutten	Elizabeth	National Science Foundation
VanSlyke	Eric	Allegany County Public Schools
Veit	Steve	Measured Progress
Verley	Jim	Wyoming Department of Education
Vitale	Michael	East Carolina University
Walqui	Aida	WestEd
Warren	Tanya	International Newcomer Academy
Weller	Mary	Howard County Public School System
Willard	Ted	National Science Teachers Association
Windschitl	Mark	University of Washington
Wood	Justine	Brandywine Springs School

C

BIOGRAPHICAL SUMMARIES OF STEERING COMMITTEE MEMBERS AND WORKSHOP SPEAKERS

Juan-Carlos Aguilar (*planning committee member*) is the science program manager for the Georgia Department of Education. He oversees state policy in the area of science education, coordinates K-12 science curriculum development, codirects Georgia's K-12 STEM initiative, supervises the alignment of the state assessments with the Georgia Performance Standards for science and serves as liaison between the Georgia Department of Education and different science organizations and institutions across the state. Previously, he was a middle school science and mathematics teacher in Fayette County. He is the president of the Council of State Science Supervisors.

Jacqueline Barber is the coprincipal investigator for *Seeds of Science/Roots of Reading,* and associate director of the Lawrence Hall of Science (LHS) in California, leading the LHS Center for Curriculum Development and Implementation and the GEMS Program. She has worked in K-12 science and mathematics education for more than 25 years, in curriculum and professional development, and is the author of many inquiry science teacher's guides. She also has a background in scientific research.

Lindsay Berk teaches science at Chinook Middle School in SeaTac, Washington. Prior to teaching middle school, she taught high school and college-level courses in science. Berk has also worked as a wildlife biologist and geographical information systems analyst for several government agencies, including the U.S. Department of Agriculture Forest Service.

Willard Brown is a professional development associate for the Strategic Literacy Initiative at WestEd. His focus is science literacy including teacher professional development, facilitator professional development, materials development, and design-based research with science teachers. Previously, Brown taught high school science in the Oakland Public Schools.

Elizabeth (Betsy) Davis is an associate professor at the University of Michigan School of Education. Her research integrates aspects of science education, teacher education, and the learning sciences. One major focus of Davis' work is a National Science Foundation (NSF)-funded project, Elementary Educative Curricula for Teachers of Science. Other projects have included the NSF-funded Curriculum Access System for Elementary Science project. Her experience also includes developing curriculum materials and serving as a teaching assistant for middle school science classes.

David Evans is the executive director of the National Science Teachers Association (NSTA). Prior to joining NSTA, he held various positions in science-related fields including director of the Center for Sustainability: Earth, Energy, and Climate at Noblis, Inc. and undersecretary for science at the Smithsonian Institution in Washington, DC. Earlier in his career, Evans was a tenured professor of oceanography at the University of Rhode Island and a classroom teacher in Media, Pennsylvania.

Susan R. Goldman is codirector of University of Illinois at Chicago's (UIC) Learning Sciences Research Institute, and a distinguished professor of psychology and education in UIC's College of Liberal Arts and Sciences. She conducts research on subject matter learning, instruction, assessment, and on roles for technology, especially in literacy and mathematics. Goldman is a board member and past president of the International Society of the Learning Sciences.

Cynthia Greenleaf is codirector of the Strategic Literacy Initiative at WestEd, where for two decades she has conducted research in adolescent literacy and translated it into teacher professional development. Currently, Greenleaf directs and serves as coprincipal investigator of *Project READi*. Concurrently, she codirects the five-year RAISE project that is bringing Reading Apprenticeship to 400,000 high school students in five states.

Monica Ko is a recent graduate of the Learning Sciences Program at Northwestern University and currently works as a visiting research specialist at the Learning Sciences Research Institute at the University of Illinois at Chicago. Her research focuses on engaging in collaborative design work to support disciplinary literacy and argumentation in K-12 science classrooms. Ko is a former high school science teacher.

Okhee Lee is a professor in the Steinhardt School of Culture, Education, and Human Development at New York University. Her current research involves the scale-up of a model of a curricular and teacher professional development intervention to promote science learning and language development of English language learners. She is a member of the writing team to develop the Next Generation Science Standards through Achieve, Inc. and the steering committee for the Understanding Language Initiative at Stanford University.

Jocelyn Lloyd is a first-grade teacher at Woodland Academy, in Worcester Public Schools, Massachusetts. She has taught at multiple grade levels for more than four years, including a combination grade 5-6 classroom in rural Maine, a science-focused classroom covering grades 5-8, and most recently a first-grade classroom in an urban district of Worcester.

Sarah Michaels (*planning committee member*) is professor of education and the senior research scholar at the Hiatt Center for Urban Education at Clark University. Her research focuses on academically productive talk in mathematics, science, and English language arts, from prekindergarten through high school. She is a coauthor of *Ready, Set, SCIENCE!: Putting Research to Work in the K-8 Science Classroom*, as well as *Accountable Talk: Classroom Conversation that Works*. She helped develop the Next Generation Science Exemplar System.

Danielle Miller is a founding faculty member of New Visions Charter High School for Advanced Math and Science in the Bronx, where she is currently a lead science teacher. In this role, she is teaching ninth-grade living environment and eleventh-grade advanced placement biology, as well as coaching department members and mentoring a Hunter College resident.

Elizabeth Birr Moje (*planning committee member*) is the associate dean for research and an Arthur F. Thurnau professor in the School of Education at the

University of Michigan. She also serves as a faculty associate in the university's Institute for Social Research, in Latino/a Studies, and in the joint program in English and education.

Jean Moon is a scholar and researcher on issues of learning and organizational and knowledge structures in education. She is president of the Tidemark Institute and a visiting scholar at the Mosakowski Institute for Public Enterprise at Clark University. She has been a principal investigator on more than 75 funded projects, particularly projects in science and mathematics education focused on professional development of teachers and college faculty and the use of technology in the delivery and learning of these subject domains.

Brett Moulding is director of the Utah Partnership for Effective Science Teaching and Learning. He was the state science education specialist, coordinator of curriculum, and director of curriculum and instruction before retiring in 2008. He taught high school chemistry for 20 years. He was a member of the National Research Council's Board on Science Education and a member of the committee who authored *A Framework for K-12 Science Education*. He subsequently served as a writing team leader for Achieve's Next Generation Science Standards.

Catherine O'Connor is professor of education and linguistics at Boston University and is currently associate dean for faculty affairs in the School of Education. She has studied classroom discussion and academically productive talk by teachers and students for more than 20 years. She has focused on the role of talk in promoting student reasoning in literacy and mathematics learning in a variety of school settings.

Jonathan Osborne is the Shriram family professor of science in the Graduate School of Education, Stanford University. He began his career teaching physics in London secondary schools before becoming a lecturer at King's College London, where he worked for 23 years as professor and department head. He joined Stanford in 2009. He was a member of the National Academy of Sciences' committee that produced the framework for the Next Generation Science Standards. Currently he is chair of the expert group responsible for producing the framework for the OECD Program for International Student Assessment science assessments in 2015.

Annemarie Sullivan Palincsar is a professor of educational studies at the University of Michigan. Her research focuses on the design of learning environments that support self-regulation in learning activity, especially for children who experience difficulty learning in school. She has also studied the role of computer-assisted instruction in enhancing children's understanding of subject-matter text and Web-based text.

P. David Pearson (*planning committee chair*) is a faculty member in the programs in language and literacy and human development at the Graduate School of Education at the University of California, Berkeley, where he served as dean from 2001-2010. He is currently working on *Seeds of Science/Roots of Reading*. He is also participating in the Strategic Education Research Partnership, a collaboration designed to embed research within the portfolio of school-based issues and priorities. Prior to coming to Berkeley in 2001, he served on the faculties of education at Michigan State University, the University of Illinois, and the University of Minnesota.

Deborah Peek-Brown is a science educator experienced in developing science curriculum and planning professional development programs for research projects at the University of Michigan. She coordinates data collection and communication with research project participants in public, private, and charter schools across the country. In addition, Peek-Brown works with the University of Michigan's teacher education program supporting student teachers, mentor teachers, and novice teachers in multiple Detroit schools. She was a science educator in Detroit public schools for 28 years.

Susan Pimentel (*planning committee member*) is a founding principal of Student Achievement Partners. She was a contributing author of the *Common Core State Standards for English Language Arts/Literacy*. Pimentel also has led several national improvement efforts, including two multistate adult education reform initiatives, and the development of content for the American Board for Certification of Teacher Excellence. Since 2007, she has served on the National Assessment Governing Board.

Kiran Purohit started her career in education as a science, mathematics, and literacy teacher at the middle school level in New York City. For the past eight years, she has worked with schools across the city as an instructional coach and curriculum developer in secondary mathematics, science, and social studies. Currently, Purohit works as the science instructional specialist for charter schools managed by New Visions for Public Schools.

Helen Quinn (*planning committee member*) is professor emerita in the Department of Particle Physics and Astrophysics at the SLAC National Accelerator Laboratory, and cochair of Stanford University's K12 Initiative. Quinn is a theoretical physicist who was inducted into the National Academy of Sciences in 2003. Her interests in education range from science curriculum and standards to the preparation and continuing education of science teachers. She was an active contributor to the California State Science Standards development. Her current Natonal Research Council committee work includes chairing the Board on Science Education and serving on the Committee on a Framework for Assessment of Science Proficiency in K-12.

Brian J. Reiser is professor of learning sciences in the School of Education and Social Policy at Northwestern University. He leads the Scientific Practices project to develop an empirically-based learning progression for scientific practices. He is also on the leadership team for Investigating and Questioning our World through Science and Technology and led Biology Guided Inquiry Learning Environments. He was a founding member and chair of the first graduate program in learning sciences, created at Northwestern University.

Sherrie Roland Grafton is a teacher at Village Elementary School in Stafford County, Virginia.

Nancy Romance is professor of science education at Florida Atlantic University where she directs a National Science Foundation project focused on researching the impact of an interdisciplinary K-2 instructional model (*Science IDEAS*). She has coauthored two elementary science textbooks series and has served as a reading consultant to several publishers of middle and high school science textbook series. A former elementary, middle and high school teacher, she was assistant principal and director of K-12 science for a large school district in southeast Florida.

Mary Schleppegrell is professor of education at the University of Michigan and chair of the Educational Studies Program. A linguist, her research studies the role of language in learning with particular attention to the needs of English language learners. She is currently engaged in an IES-funded project, *Exploring Language and Meaning in Text with English Language Learners.*

Sam Shaw has served as the state science education specialist for South Dakota for over three years. He has consulted for and helped develop many initiatives involving teacher training to build instructional capacity. Shaw also led South Dakota's Next Generation Science Standards work as a lead state and is currently working through a final review of those standards. His previous experience includes teaching middle school science.

LeeAnn M. Sutherland is an associate research scientist in the School of Education at the University of Michigan. She joined the faculty in 1991, teaching in what is now the Sweetland Writing Center in the School of Literature, Science, and the Arts before obtaining her doctorate and joining the School of Education faculty. She is coprimary investigator of *Investigating and Questioning our World through Science and Technology*. Certified as an English teacher at the secondary level, Sutherland has worked in rural, urban, and suburban middle and high schools.

Michael Vitale is a professor of curriculum and instruction at East Carolina University. He has worked in both K-12 school research and university settings. His school research experience includes serving as director of Applied Research and as director of Instructional Technology for the Dallas (Texas) Independent School District. His university experience includes serving as an assistant professor of educational psychology at the University of Hawaii, associate professor and coordinator at the Center for Educational Technology at Florida Atlantic University, and codirector of the Educational Research Laboratory.

Aida Walqui is the director of the Teacher Professional Development Program at WestEd. Previously, she taught in the Division of Education at the University of California, Santa Cruz, and the School of Education at Stanford University. She has also taught in other universities in England, Mexico, Peru, and the United States.

Mark Windschitl is a professor of science teaching and learning at the University of Washington. His research interests deal with the early career development of science teachers. He has recently been principal investigator (PI) on two projects that tracked science teachers from preparation through their first year of teaching. His research group has prototyped a set of high-leverage practices for K-12 science instruction that represent a "beginner's repertoire" and has tested the conditions under novices can apply these core practices. Windschitl is also PI on a Noyce Teaching Scholars grant at the University of Washington.